THE BOVINE SCOURGE

Meat, Tuberculosis and Public Health, 1850–1914

By the late 1890s, the question of bovine tuberculosis (TB) and infected meat had become one of national importance, reflecting a national sense of fear. Although the extent of the threat to health proved uncertain, bovine TB had come to stand at the centre of debates about diseased meat and public health. The anxiety it caused was part of a longer story, linked to concern over food safety, changes in how tuberculosis was understood, and worries over diseased meat and the 'evils' of the urban meat trade. *The Bovine Scourge* explores the debates and fears that came to surround bovine TB, meat and public health between the 1860s and 1914. It traces how diseased meat and bovine TB emerged as a public health issue, examines the measures adopted to protect the public, and addresses how by the Edwardian era milk had become the major source of concern in discussion of bovine TB. It also raises important questions about the history of food safety, the concerns generated by diseased meat, and the role of the public health and veterinary profession in preventing the sale of contaminated food.

Dr Keir Waddington is Senior Lecturer in History at the University of Cardiff.

THE BOVINE SCOURGE

Meat, Tuberculosis and Public Health, 1850–1914

Keir Waddington

THE BOYDELL PRESS

First published 2006
The Boydell Press, Woodbridge

ISBN 1 84383 193 7

The Boydell Press is an imprint of Boydell & Brewer Ltd
PO Box 9, Woodbridge, Suffolk IP12 3DF, UK
and of Boydell & Brewer Inc.
668 Mt Hope Avenue, Rochester, NY 14620, USA
website: www.boydellandbrewer.com

A CiP catalogue record for this book is available
from the British Library

Typeset by Keystroke, Jacaranda Lodge, Wolverhampton
Printed in Great Britain by
Cambridge University Press

Contents

Acknowledgements vii
List of abbreviations ix

1 Introduction 1

2 Diseased meat and public health, 1850–1870 11

3 Coughing cows: bovine tuberculosis and contagion 30

4 'No inconsiderable danger to the community': meat and bovine tuberculosis 52

5 Inspecting meat 70

6 Making of an epoch? 92

7 Experimentation and the British model 112

8 Diseased meat control and the 'great modern scourge' 131

9 Death in the milk pail: tuberculosis and milk 153

10 Eradication? Meat, milk and bovine tuberculosis 175

Select bibliography 192
Index 209

Acknowledgements

The research for this book would never have been possible without the luxury of a Cardiff Research Fellowship or the encouragement and interest shown by Anne Hardy and the enthusiasm exhibited at the outset by Jennifer Haynes. Nor could I have completed the book without the assistance and encouragement of Faye Hammill and Rose Thompson who were always prepared to ask questions and read drafts.

In researching this book I want to thank Vivienne Aldous, Jessica Newton, and James Sewell at the Corporation of London Record Office for their help and patient assistance. Thanks are owed to Kerstiene Lang and Rachel Taylor at the Glasgow City Archive, Mitchell Library, for their kindness and aid. I would also like to thank the staff at the Glamorgan Record Office, London Metropolitan Archive, and the National Archive for their assistance. I am grateful to the staff of the British Library, British Newspaper Library at Colindale, Bodleian Library, Cardiff Arts and Social Sciences Library, and as ever to the staff at the Wellcome Library for their invaluable help and constant good cheer.

Considerable assistance has also been provided by numerous colleagues and friends. A great debt is owed to those whose careful reading and many useful comments have been invaluable, and in particular to Roberta Bivins, Bill Jones, Christopher Marlow and Richard Sugg for their patient reading and interest. I am grateful to Peter Atkins, Roger Cooter, Gareth Enticott, Elsbeth Heaman, Martin Daunton, Pat Hudson, Susan Jones, John Pickstone, the late Roy Porter, Frank Trentmann, Garthine Walker, Lise Wilkinson, Abigail Woods and Mick Worboys for their suggestions, questions and encouragement. Gratitude is also owed to Peter Sowden at Boydell for making this book possible and for his support throughout the publication process. Any mistakes are of course my own.

Keir Waddington
Cardiff University, September 2005

Abbreviations

AHR	*Agricultural History Review*
BHM	*Bulletin of the History of Medicine*
BMA	British Medical Association
BMJ	*British Medical Journal*
BSE	Bovine spongiform encephalopathy
CJD	Creutzfeldt-Jakob Disease
CLRO	Corporation of London Record Office
DPH	Diploma of Public Health
EcHR	*Economic History Review*
EMJ	*Edinburgh Medical Journal*
EVR	*Edinburgh Veterinary Review*
EVJ	*Edinburgh Veterinary Journal*
GCA	Glasgow City Archive, Mitchell Library
GMC	General Medical Council
GMJ	*Glasgow Medical Journal*
GRO	Glamorgan Record Office
JCPT	*Journal of Comparative Pathology and Therapeutics*
JEcH	*Journal of Economic History*
JRAS	*Journal of the Royal Agricultural Society*
JSM	*Journal of State Medicine*
LCC	London County Council
LGB	Local Government Board
LMA	London Metropolitan Archive
MAMOH	Metropolitan Association of Medical Officers of Health
MOH	Medical officer of health
MOsH	Medical officers of health
MPC	*Medical Press and Circular*
MRC	Medical Research Council
MTG	*Medical Times and Gazette*
MTJCSG	*Meat Trades' Journal and Cattle Salesman's Gazette*
PP	Parliamentary Papers
PRO	Public Record Office
PRSM	*Proceedings of the Royal Society of Medicine*
SHM	*Social History of Medicine*
SMOH	Society of Medical Officers of Health
TNA	The National Archive
RCVS	Royal College of Veterinary Surgeons
RVC	Royal Veterinary College

1

Introduction

The emergence of food scares in the 1980s and 1990s, first around salmonella and then bovine spongiform encephalopathy (BSE), are dramatic examples of the growing consumer and media alarm that came to surround food safety and the relationship between animal and human disease in the late twentieth century.[1] In the process, consumers became more sensitive to the risks associated with food and began to raise questions about who was responsible for ensuring food safety. Although the 1988 crisis over salmonella in eggs has been considered the first food scare since the 1964 Aberdeen outbreak of typhoid in tinned corned beef, it was BSE that made food safety a serious public and political issue, forcing the subject of meat from diseased cattle onto the agenda.

When BSE was first diagnosed in 1986 veterinarians and scientists were initially puzzled. The nature of the infectious agents causing transmissible spongiform encephalopathies was a matter of controversy, but a link was made between BSE and a family of brain diseases that included Creutzfeldt-Jakob Disease (CJD). Although nobody knew whether BSE was a hazard to human health, because the two diseases were believed to come from the same family the possibility that BSE could be passed to man through meat created alarm. The media fuelled hysteria about the risk to human health and as a sense of crisis developed controversy raged over whether a human epidemic of BSE was possible, and who was most at risk. Although incidences of BSE in cattle had fallen by the mid 1990s, the growing number of deaths from CJD among farmers with BSE infected herds suggested a link between the two and that beef from diseased cows was dangerous. Studies in 1994 had already explained that BSE might be a risk, but it was not until two years later that the government was advised that the most likely cause of this new variant of CJD was eating beef products contaminated with BSE. Repeated government assurances that beef was safe appeared to be disastrously unfounded. As the number of cases of vCJD rose, so too did public fears about British beef.

The belated measures put in place to protect the public from the risk of BSE did little to mollify alarm about food safety. Beef, one of the essential constitutes of British national identity for the political right, became a target for national anxiety and a source of international concern. With the public

[1] See D. Miller and J. Reilly, 'Making an Issue of Food Safety', in *Eating Agendas: Food and Nutrition as a Social Problem*, ed. D. Mauer and J. Sobal (New York, 1995), pp. 305–36.

now sensitive to the potential dangers posed by animal diseases and meat, reports of any cattle disease that could cross the species barrier attracted considerable media attention. The media was quickly rewarded with another potential food scare: just as the furore surrounding BSE began to subside, reports appeared in 1997 that levels of bovine tuberculosis in cattle had risen sharply. Bovine tuberculosis was not a new disease, however. The pathways of infection through meat and milk had been established in the late nineteenth century and an active eradication policy from the 1950s onwards had seen the disease all but stamped out. Although the public were reassured that there was no immediate threat, Britain was at risk of losing its bovine tuberculosis free status. Farmers were once more blamed for failing to control a cattle disease and were attacked for supporting measures – mainly the cull of badgers – that had little scientific rationale or proven efficacy. By 2004, the Department of the Environment, Food and Rural Affairs – the successor to the Ministry of Agriculture, Fisheries and Food – considered bovine tuberculosis as 'the largest threat facing us at the moment'.[2] Although there was no reported rise in infection from the bovine form of the disease in the human population, because the increase in bovine tuberculosis in cattle followed on so closely from BSE the possible danger to public health was immediately investigated.

Concerns about the relationship between bovine tuberculosis and food safety are not unique to the late twentieth century, however. Unease about food and public health has a long history. The relationship between the two has been an intermittent object of concern since the start of the nineteenth century, although the parameters of this anxiety have shifted from a desire to protect the public from fraud to alarm about the role of meat and milk in the transmission of disease. Historical interest in nineteenth century debates on adulteration has, however, tended to mask the Victorians' growing unease about the safety of meat.[3] At first an ill-defined threat, by the 1870s concerns about the health risks of eating diseased meat had come to focus on tuberculous meat. Contemporary fears of the resurgence of bovine tuberculosis and the numerous recommendations to control its spread echo Victorian and Edwardian responses to bovine tuberculosis.

A Victorian BSE?

The coverage of the BSE crisis has raised serious questions about food safety, the way science is employed, and the nature of the farming industry, but if these concerns appear topical and ripe for alarmist media coverage now, our current anxieties about infected meat are not new. The threat of animal

[2] DEFRA, *Preparing for a new GB Strategy on Bovine Tuberculosis* (London, 2004), p. 8.
[3] See John Burnett, *Plenty and Want: A Social History of Diet in England from 1815 to the Present Day* (London, 1966).

diseases was all too familiar to the Victorian and Edwardian public and was widely reported in the press. Rabies prompted hysteria, whilst cattle epizootics raised the spectre of contagion and generated alarm about the safety of meat from infected livestock. A close relationship has always existed between animal and human disease, but the increasing domestication and exploitation of animals in the nineteenth century created new opportunities for the spread of disease from animals to humans, whilst the growing urban consumption of meat and milk provided an effective path for the transfer of zoonotic diseases. The very visibility of epizootic and endemic cattle diseases served to increase apprehension about the threat to human health from diseased meat, encouraging investigation into the diseases of animals and their relationship with disease in humans. Evidence of the evils of the meat trade and the proliferation of local food scares contributed to a growing sense of alarm about food safety and meat.

At first evidence of the direct transmission of animal diseases through meat was mainly anecdotal. However, studies into bovine tuberculosis from the 1860s onwards appeared to offer proof that one of the most widespread diseases in cattle could be passed to humans. The contagious properties of bovine tuberculosis and its ability to cross the species barrier were confirmed by bacteriological studies in the 1880s, which established clear links between animal and human disease. Bovine tuberculosis made diseased meat into a concrete threat. It pushed the issue of diseased meat into the political arena and quickly came to stand at the heart of Victorian and Edwardian debates on animal disease and public health. Anxiety focused on the domestic meat supply; little unease was expressed about canned, frozen or preserved meat, which was generally considered free from infection, at least until the revelations about the Chicago meat trade in 1907 directed attention onto imported meat. Conflicting claims about the threat from infected beef were made, but the main impression was that meat exhibiting signs of tuberculosis was a danger to consumers. By the 1890s, bovine tuberculosis had become the model zoonosis – in much the same way as BSE was to over a century later – and reshaped the concept of safe food, so that it was food as an agent of disease and not just concerns about adulteration and contamination that now worried the medical and veterinary professions. The alarm generated by bovine tuberculosis merged with existing disquiet about the condition of urban cowsheds and their impact on animal and human health, fears of national degeneration, and anxieties about child health to create a potent mix.

Concern was not limited to Britain. Fears about the threat from tuberculous meat were international and the responses to the disease followed a similar pattern, if not the same timescale, in Europe and North America.[4] By 1900,

[4] See J. Antunes *et al.*, 'Tuberculose e Leite', *História Ciências Saúde: Manguinhos* ix (2002), pp. 609–23; Barbara G. Rosenkrantz, 'The Trouble with Bovine Tuberculosis', *BHM* lix (1985), pp. 155–75; Peter Koolmees, 'Veterinary Inspection and Food Hygiene in the

bovine tuberculosis had become part of the international fight against tuberculosis and a focal point for intellectual debates on the production of disease in humans through the agency of food. In Britain, existing public health measures to prohibit the sale of unwholesome meat were tightened. Import controls were introduced. Experimental studies were commissioned into the extent of the danger and guidelines were belatedly issued on what meat from infected cattle was safe to eat. Campaigns were launched to eradicate the disease in cattle and research was conducted into a possible vaccine. In the light of BSE all this sounds strangely familiar. But was bovine tuberculosis the Victorian equivalent to BSE?

At a superficial level, there are resonances between late twentieth century responses to BSE and Victorian debates on bovine tuberculosis. In both cases, their zoonotic properties made a cattle disease into a public health issue and focused existing unease about food and disease onto meat. Further parallels can be found. The nature and route of infection in BSE and bovine tuberculosis were initially open to considerable speculation. Only gradually did experimental studies confirm established prejudices that meat was a possible agent of infection. In both cases, the state turned to science to determine how safe meat from infected livestock was, and to buy time to play down the risks. Similar problems were encountered in identifying diseased carcasses. Confusion existed over what to inspect for and what should be removed to render carcasses safe. Accusations were made that the state seemed more concerned to protect farming interests than public health, whilst tensions between public health and agricultural concerns led government departments to shift responsibility. In addition, BSE and bovine tuberculosis stimulated debate on the relationship between veterinary and human medicine.

However, it would be unwise to draw too many parallels. Although BSE and bovine tuberculosis provoked intense debate about the threat posed by diseased meat, there are important differences in how the two diseases were constructed, the way in which science was used, the role the media played, and in the level of public and professional concern. No attempt was made to define bovine tuberculosis as an animal health or agricultural issue. From the start, it was perceived as a public health problem and formed part of a network of food scares which, according to French and Phillips, characterised the evolution of British food laws.[5] Yet, despite these important differences, given contemporary unease about rising levels of bovine tuberculosis and fears of diseased meat, a study of the emergence of bovine tuberculosis as a zoonosis and the public health responses reveals how unease about diseased meat has

Twentieth Century', in *Food, Science, Policy and Regulation in the Twentieth Century: International and Comparative Perspectives*, ed. David F. Smith and Jim Phillips (London, 2000), pp. 53–68.

[5] Michael French and Jim Phillips, *Cheated Not Poisoned? Food Regulation in the United Kingdom, 1875–1938* (Manchester, 2000).

a long history, and how fears of the disease had an important bearing on Victorian and Edwardian responses to the regulation of meat and milk. As one witness to the Environment, Food and Rural Affairs Committee noted in 2004, 'I think we need to look back in history and re-examine why we started dealing with bovine TB as a disease.'[6]

The bovine scourge

Although tuberculosis has been central to debates on nineteenth century improvements in health and the role of social intervention in mortality change, limited scholarly attention has been directed at the bovine form of the disease. The emergence of multi-resistant strains of tuberculosis in recent years has encouraged a re-evaluation of the history of the disease, whilst national and local studies have added a new layer of complexity in understanding responses to tuberculosis.[7] However, the literature on tuberculosis has remained primarily concerned with the 'crusade against consumption', examining the iconography of the disease, sufferers' experiences, and the institutional responses to the 'white plague'. Whereas Atkins, Smith and Worboys have touched on the concerns generated by tuberculous meat, historians have generally neglected bovine tuberculosis, preferring instead to focus on the more emotive subject of pulmonary tuberculosis.[8] What attention has been directed at bovine tuberculosis has mainly concentrated on milk and child health, or the difficult adoption of pasteurisation.[9] The public health movement's battle against the

[6] House of Commons, Environment, Food and Rural Affairs Committee, Transcript of Oral Evidence, 26 May 2004.

[7] For example, Linda Bryder, *Below the Magic Mountain: A Social History of Tuberculosis in Twentieth-century Britain* (Oxford, 1988); David S. Barnes, *The Making of a Social Disease: Tuberculosis in Nineteenth Century France* (Berkeley, 1995); Katherine McCuaig, *The Weariness, the Fever, and the Fret: The Campaign against Tuberculosis in Canada, 1900–50* (Montreal, 1999); Anna Shell, 'TB or not TB eradicated?' (Cardiff University, PhD diss.). Work by Rosenkrantz and Jones does go some way to examining international responses to bovine tuberculosis but their focus is primarily on North America: Rosenkrantz, 'The Trouble with Bovine Tuberculosis', pp. 155–75; Susan Jones, 'Placing Disease: Transnational Perspectives on the Bovine Tuberculosis Debate, 1901–14', paper presented at SSHM spring conference, Sheffield Hallam University, Mar. 2002.

[8] Peter J. Atkins, 'The Glasgow Case: Meat, Disease and Regulation, 1889–1924', *AHR* lii (2004), pp. 161–82; F. B. Smith, *The Retreat of Tuberculosis, 1850–1950* (London, 1988), pp. 175–94; Michael Worboys, *Spreading Germs: Disease Theories and Medical Practice in Britain, 1865–1900* (Cambridge, 2000), ch. 7.

[9] See Peter J. Atkins, 'White Poison? The Social Consequences of Milk Consumption, 1850–1930', *SHM* v (1992), pp. 207–27; Deborah Dwork, 'The Milk Option: An Aspect of the History of the Infant Welfare Movement in England, 1898–1908', *Medical History* xxxi (1987), pp. 51–69; Peter J. Atkins, 'The Pasteurisation of England: The Science, Culture and Health Implications of Milk Processing, 1900–50', in *Food, Science, Policy and*

sale of tuberculous meat in its fight against tuberculosis has therefore been neglected. So too has the study of public health in relation to animals. Although recent studies have begun to explore the relationship between food and public health, few historians have examined the links between animal and human disease.[10] Similarly, work on the regulation of food has also tended to underplay its role in disease causation.[11] In examining the public health concerns generated by tuberculous meat between 1850 and the outbreak of war in 1914, this study goes some way to addressing these gaps in the literature. Although how bovine tuberculosis came to be understood (or 'constructed') as a zoonotic disease forms a central part of this study, it offers a contextual examination of how bovine tuberculosis shaped the understanding of diseased meat and milk and made them into a public health issue. But how did bovine tuberculosis come to be perceived as a danger to public health?

Part of the answer lies in a shift in concerns about meat in the 1840s and 1850s from an awareness that unwholesome meat was 'unfit for human consumption' to a sense that flesh from diseased livestock might be prejudicial to health. A correlation existed between rising meat consumption and the emergence of debate on the dangers of eating diseased meat. Historians have tended to examine meat consumption as a component of improvements in the standard of living, but as chapter 2 explains, little attention has been directed at the quality of meat purchased. As evidence of the evils of the meat trade accumulated, contemporaries became worried that large quantities of meat from diseased livestock were being sold to the urban poor. It was against this background that medical officers of health (MOsH) began to warn of the dangers of eating diseased meat.

Although a connection between tuberculosis in animals and humans had long been suspected, it was not until the 1860s that what might be labelled scientific interest was directed at the issue. As chapter 3 demonstrates, European studies into cattle diseases and microbes began to look at the association between tuberculosis in cattle and the disease in humans as other cattle diseases were dismissed as posing little danger to health. Pathologists were turning to the study of contagious animal diseases in an attempt to find answers to the way disease spread. The idea that bovine tuberculosis was

Regulation in the Twentieth Century, ed. Smith and Phillips, pp. 37–51. Perren, despite his interest in how the meat trade was regulated, paid little attention to bovine tuberculosis: Richard Perren, *The Meat Trade in Britain, 1840–1914* (London, 1978), pp. 89–90, 134–6.

[10] Anne Hardy, 'Animals, Disease and Man: Making Connections', *Perspectives in Biology and Medicine* xlvi (2003), pp. 200–15; Lise Wilkinson, *Animals and Disease: An Introduction to the History of Comparative Medicine* (Cambridge, 1992); *idem*, 'Zoonoses and the Development of Concepts of Contagion and Infection', in *The Advancement of Veterinary Science: The Bicentenary Symposium Series*, vol. iii, *History of the Health Professions: Parallels between Veterinary and Medical History*, ed. A. R. Michell (Oxford, 1993), pp. 73–90.

[11] For a survey of the recent literature, see John C. Super, 'Food and History', *Journal of Social History* xxxvi (2002), pp. 165–78.

contagious and capable of crossing the species barrier was therefore almost a by-product of investigations into the pathology of contagion and the growth of comparative medicine as a discipline. Interest in Britain, however, was generated initially by the veterinary profession and was not determined, as some historians have suggested, by MOsH. The contribution of veterinary medicine to medical ideas and public health remains relatively unexplored and chapter 3 shows how it was veterinarians who first defined bovine tuberculosis as a public health risk. Doctors only started to become interested in the disease in the mid 1870s. Attention was fuelled by discussions about the communicable nature of tuberculosis and mounting anxiety about food safety. The different interpretations adopted by veterinarians and doctors emphasise the uneven spread of medical ideas between countries and between practitioners.

Yet, despite growing support in veterinary and medical circles for the idea that tuberculosis was inter-communicable, it was Robert Koch's claim in 1882 that bovine and human tuberculosis were caused by the same organism that was felt to offer conclusive proof. However, the impact of Koch's ideas on bovine tuberculosis needs to be reassessed in the same way as his work on pulmonary tuberculosis and germ theory has been re-evaluated. As chapters 3 and 4 illustrate, although Koch's identification of the tubercle bacilli asserted the value of bacteriology and the idea of contagion in medical discourse, investigations in Europe had already pointed to the suggestive likeness between tuberculosis in humans and cattle and the pathogenic properties of meat from infected livestock. What Koch did do was stimulate medical interest in bovine tuberculosis. Pathological and bacteriological studies acquired an important role in defining and defending the idea that bovine tuberculosis was a zoonosis, suggesting that pathological and laboratory acquired knowledge could be quickly incorporated by veterinarians and sanitarians even if the 'effective infiltration of laboratory science into public health knowledge and practice' occurred only gradually in other areas.[12]

Claims that bovine tuberculosis was inter-transmissible were translated into alarm about meat's role in spreading the disease. However, as chapter 4 shows, whereas many veterinarians and MOsH by the 1880s could accept the idea that bovine tuberculosis could cross the species barrier and that meat was an agent of infection, the exact danger from meat was harder to determine. Once more, differences of interpretation were adopted by veterinarians and MOsH as each tried to assert their own expertise. The resulting confusion not only highlights inter-professional conflict but also frustrated efforts to condemn tuberculous carcasses, pointing to the problems conflicting interpretations created in public health.

[12] Anne Hardy, 'Priorities in the History of Public Health and Preventive Medicine to 1945', in *Public Health and Preventive Medicine, 1800–2000: Knowledge, Co-operation and Conflict*, ed. Astri Andresen, Kari Tove Elvbakken and William H. Hubbard (Bergen, 2004), p. 18.

The difficulties caused by these differing interpretations came to a head in Glasgow in 1889 when attempts were made to prosecute a butcher and salesman for selling tuberculous meat. The trial focused national debate and revealed the failure of experts to agree over the nature of bovine tuberculosis and the threat from meat. To clear up this confusion, a royal commission was appointed. It confirmed that expert opinion was so divided that it could not be relied upon to shape policy. Instead, the commissioners turned to a programme of experimental research. Chapter 6 reinforces the idea that if the state was a reluctant patron it was prepared to fund research into areas it considered of national importance. The commission fashioned a British model that bovine tuberculosis was a hazard to public health, although a second commission was required to outline guidelines on how best to deal with tuberculous meat. So entrenched were these ideas that when Koch challenged this model of tuberculosis in 1901 there was an immediate backlash and a further royal commission was appointed. As chapter 7 explains, from the start the commissioners used the 'objective' value assigned to science to refute Koch's assessment, a finding that continued to shape responses to tuberculous meat in the first half of the twentieth century. However, much of the work proved to be more like cooking than science, raising wider questions about the nature of scientific research in Edwardian Britain.

Despite problems in defining the nature of the threat, measures to combat the spread of bovine tuberculosis concentrated on limiting human infection through the regulation of the meat and milk trade. Prevention through a system of surveillance and meat inspection, first as part of the broader ideology of nuisances that existed in the mid Victorian public health movement and then through direct attempts to regulate the meat trade, was crucial to attempts to protect consumers who were often cast as innocent dupes. The nature of this meat inspection is examined in chapters 5 and 8. These chapters concentrate less on legislation or the neat chronologies found in some studies of public health to look instead at how inspection was organised, relating this to questions of expertise and identification. Although by the end of the nineteenth century public health had become 'the most powerful justification for urban public action', the process was far from linear.[13] An examination of how meat inspection was organised at a local level raises questions about the nature of the public health movement and the problems encountered. Intervention remained uneven, and, as Hamlin has shown in other areas of public health, there were important differences between towns.[14] Marked variations also

[13] Christopher Hamlin, 'Public Sphere to Public Health: The Transformation of "nuisance"', in *Medicine, Health and the Public Sphere, 1600–2000*, ed. Steve Sturdy (London, 2002), p. 191.

[14] Christopher Hamlin, 'Muddling in Bumbledom: On the Enormity of Large Sanitary Improvements in Four British Towns, 1855–85', *Victorian Studies* xxxii (1988/9), pp. 55–83.

existed between urban and rural responses, an area of inquiry that has been neglected in histories of public health.

Whereas many studies of urban public health have emphasised the important role MOsH had in shaping local sanitary policy, chapters 5 and 8 demonstrate how much of the day-to-day work was carried out by subordinate officials who had their own professional worries. These officials have often been neglected by historians.[15] By examining the work of these subordinate officials further insights into the limitations of local public health work can be gained. MOsH often adopted a supervisory role and their authority to determine what meat should be seized was contested not only by butchers but also by magistrates and veterinarians. Although doctors asserted that tuberculous meat was their field of responsibility, veterinarians made similar claims to competence. Tensions therefore emerged between veterinary surgeons and doctors not only over the definition of the disease, as discussed in chapter 3, but also over who should be responsible for meat inspection.

Meat inspection remained chaotic. Improvement was slow and by the end of the nineteenth century the problems that had dogged meat inspection since the 1860s continued to be unresolved. Victorian and Edwardian public health, it would seem, were not always about the successful elimination of contagious diseases or improvement, whilst the regulation of food remained problematic. Attempts to reform the meat trade came into conflict with trade interests, highlighting the tensions between the public and commercial spheres. French and Phillips have shown how commercial interests responded to and shaped food regulation, and how different pressure groups 'used their own notions of the "public interest" as convenient rhetoric in the debates about food regulation'.[16] This is clear in moves to prevent the sale of diseased meat.

By 1900 the alarm surrounding meat had started to be replaced by concerns about milk. Chapter 9 explores how by 1914 milk had become the principal focus of debate on bovine tuberculosis. Although more historical attention has been directed at the problem of milk, it does need to be re-examined in the context of the public health movement and not just in terms of child health or pasteurisation. In looking at these issues, the chapter returns to earlier themes by examining the tensions between doctors and veterinarians, how trade interests frustrated reform, and how the reluctance of rural authorities to act created problems. Bovine tuberculosis remained the hidden enemy in the milk pail.

The failure to regulate the meat and milk trade recast debate in the Edwardian period to focus on stamping out the disease in cattle. Throughout

[15] See Dorothy E. Watkins, 'The English Revolution in Social Medicine, 1889–1911' (unpubl. PhD diss., London, 1984); Anne Hardy, 'Public Health and Experts: The London Medical Officers of Health', in *Government and Expertise: Specialists, Administrators and Professionals, 1860–1919*, ed. Roy MacLeod (Cambridge, 1988), pp. 128–42.
[16] French and Phillips, *Cheated Not Poisoned?* p. 191.

the interwar period, eradication and pasteurisation became the preferred solutions to removing the threat from bovine tuberculosis, but the mechanisms established to regulate the meat and milk trade in the late nineteenth and early twentieth centuries continued to shape efforts to prevent the spread of animal diseases to humans. A relaxation of this preventive framework in the 1980s and 1990s coincided with the re-emergence of food scares. In 2004, the death of a number of cows from an unknown virus – possibly cattle polio – sparked new concerns about the risk to humans from cattle. The outbreak only served to reinforce fears of the potential danger of a cattle disease to human health and the need to regulate the meat trade, concerns that were all too familiar to the Victorian and Edwardian public.

2

Diseased meat and public health, 1850–1870

Tuberculosis has a long history. It was considered a major cause of death in the seventeenth century, but from the late eighteenth century the transition to higher levels of urbanisation saw the disease reach epidemic proportions and become a major contributor to high mortality in most European countries. Although levels of infection were higher in rural districts, the disease was seen as a predominantly urban phenomenon that largely affected young adults from the poorer classes. Spread by the tubercle bacillus (*mycobacterium tuberculosis*), it could enter the body in a number of ways and hence took on a variety of forms, making it hard to pin down. Early symptoms were not easy to detect and it was only in the advanced stages that tuberculosis became obvious. Nor are statistics reliable, especially as the disease was often described euphemistically to avoid the associated stigma.[1] Despite these methodological problems in determining the incidence of tuberculosis, historians agree that mortality from the disease declined in the last quarter of the nineteenth century. It was a process shaped not only by legislative action aimed at the disease itself but also by a wide range of environmental improvements that affected the transmission of the disease.[2]

This overall trend does not tell the entire story, however. Although the crusade against consumption was primarily directed at pulmonary tuberculosis, growing anxiety was expressed about the other forms of the disease, which did not fall at the same rate. Of these, abdominal tuberculosis (or tabes mesenterica), linked to bovine tuberculosis and the ingestion of infected meat and milk, generated the most alarm.

Just as with pulmonary tuberculosis, the bovine form of the disease was considered pre-eminently a disease of the urban poor. However, because of the nature of the disease and the records, the actual number of deaths from bovine tuberculosis is hard to determine. Local variations in incidence were detected, but contemporaries broadly agreed that between 5 and 10 per cent of those dying from tuberculosis did so from its bovine form. By 1899, rough calculations from mortality returns placed the annual death rate at approximately 3,000,

[1] Linda Bryder, '"Not Always One and the Same Thing": Registration of Tuberculosis Deaths in Britain, 1900–50', *SHM* ix (1996), pp. 253–65.

[2] Gillian Cronje, 'Pulmonary Tuberculosis in England and Wales' (unpubl. PhD diss., LSE, 1990); Simon Szreter, 'The Importance of Social Intervention in Britain's Mortality Decline c.1850–1914: A Reinterpretation of the Role of Public Health', *SHM* i (1988), pp. 1–37.

with a larger number crippled.[3] What alarmed contemporaries most was that the disease appeared to be on the increase and, as susceptibility diminished with age, that children were at greater risk. If the death rate from pulmonary tuberculosis had started to fall by the end of the nineteenth century, mortality in children under 12 years from the abdominal form of the disease had increased, a trend that ran counter to the general fall in infant mortality after 1890. It was assumed that in the first decade of the twentieth century one-third of children between 5 and 16 years of age contracted bovine tuberculosis.[4] Mounting fears about national degeneration and the importance ascribed to infant life and child health in debates about the health of the nation made these statistics even more frightening.

Public health officials and veterinarians were keen to trot out these statistics to support their claims that action was needed to protect the public and above all children from this insidious disease. Although contemporaries were themselves sceptical of the statistics, noting that the danger had probably been exaggerated, they continued to use them in an increasingly alarmist fashion as evidence that bovine tuberculosis was a threat to public health. However, epidemiological trends mattered less than perceptions. Part of the reason why late Victorian and Edwardian doctors, veterinarians and tuberculosis campaigners were so alarmed about bovine tuberculosis was not just that the disease could cross the species barrier and infect humans, but that it was endemic in cattle. This, it was feared, made it difficult to escape ingesting meat and milk from infected livestock, hence increasing the potential to contract the disease.

However, the identification of 'coughing cows' was 'very often difficult or impossible', even for experienced veterinarians.[5] Only when the disease was advanced did the symptoms become acute, leading to emaciation, raised temperatures, and breathlessness. Post-mortem evidence varied, depending on the site of infection, and tuberculosis in cattle was frequently mistaken for other conditions. These problems with identification did not stop contemporaries from suspecting that levels of bovine tuberculosis were rising; only a few noted that this might be because as doctors and veterinarians had become more closely involved in meat and milk inspection after 1850 they had begun to pay more attention to the disease. Concern focused primarily on domestic supply; imported cattle were generally considered free from tuberculosis because of the healthier, open-air conditions under which they were reared. For domestic cattle, however, it was felt that the disease was endemic. Wild estimates were made as to the extent to which the national herd was affected.

[3] Arthur Littlejohn, *Meat and its Inspection: A Practical Guide for Meat Inspectors, Students, and Medical Officers of Health* (London, 1911), p. 171.

[4] William Savage, *Milk and the Public Health* (London, 1912), p. 149; *British Journal of Tuberculosis* i (1907), p. 250.

[5] Hugh Macewen, *Food Inspection: A Practical Handbook* (London, 1909), p. 31.

Writing in 1911, W. Shipley, veterinary inspector for Yarmouth, confidently claimed that 'there is not a dairy . . . in the northern and eastern districts free from tuberculosis.'[6] Others pointed to incidences of infection in the national herd of 50 per cent or more, with individual farms reporting levels of the disease as high as 65 per cent.[7] These high rates of infection were confirmed by anecdotal and impressionistic evidence that 'there can be no doubt that in all classes of cattle, Tuberculosis is now universally prevalent.'[8]

Problems with these statistics were overlooked: strong regional variations existed, with London, Liverpool and Manchester reporting much higher levels of the disease than other parts of the country. Few commented that most of the statistics referred to dairy cows who were more likely to contract the disease because of the confined and intensive conditions under which they were kept.[9] Tuberculin testing in Europe in the early twentieth century revealed more modest levels of infection than contemporaries feared, pointing to between 20 to 30 per cent of cattle with tuberculosis. A similar percentage was adopted by the 1901 London congress on tuberculosis, a figure widely repeated by Edwardian writers on veterinary medicine, dairy farming and public health.[10] However, local programmes of tuberculin testing indicated that the number of infected cattle in Britain compared unfavourably with reported levels in the rest of Europe. For example, when the Earl of Spencer's cows were tested in 1893, all but one of the twenty-three strong herd reacted; when slaughtered all were found to be tubercular. Testing at other farms revealed comparable figures. What made these statistics more alarming was that many of the cattle tested had been in apparent health. For John MacFadyean at the Royal Veterinary College (RVC) this proved that tuberculosis was more common among cows 'in this country than hitherto imagined'.[11]

Doctors and veterinarians initially turned to medical models of predisposition and atmospheric stagnation to explain these high levels of infection. In the same way as urban conditions were used to explain high levels of pulmonary tuberculosis, the unhygienic conditions under which cattle were kept were also considered important in understanding why bovine tuberculosis remained endemic, fuelling calls for improvements to dairies and cowsheds. A bacteriological model of tuberculosis following Robert Koch's identification of the tubercle bacillus in 1882 (see chapter 4) did not immediately overturn these ideas. Just as with pulmonary tuberculosis, a hereditary understanding

[6] 'Tuberculous Meat and Milk', *Sanitary Record*, 7 Dec. 1911, p. 557.

[7] See 'The Communicable Diseases Common to Man and Animals and their Relationship', *ibid.*, 15 Sept. 1888, p. 124.

[8] 'Report on Bovine Tuberculosis', *JCPT* i (1888), p. 178.

[9] Peter J. Atkins, 'The Glasgow Case: Meat, Disease and Regulation, 1889–1924', *AHR* lii (2004), p. 164.

[10] *Transactions of the British Congress on Tuberculosis* iv (1901), p. 44.

[11] *JRSA* v (1894), p. 116; John MacFadyean, 'Results of the Application of the Tuberculin Test to Her Majesty's Dairy Cows at Windsor', *JCPT* xii (1899), pp. 50–6.

of bovine tuberculosis was maintained and veterinarians continued to argue that certain breeds were inherently disposed to the disease, a view that reassured anxious elite cattle-breeders.[12] As a bacteriological view came to dominate interpretations, farmers were discouraged from thinking in hereditary terms and a further set of explanations were added. Considerable interest was shown in how the bacillus was transmitted between cattle and in the environment, with manure and hay identified as infective agents.[13] However, despite the shift in how bovine tuberculosis was perceived, methods to combat the disease in cattle continued to focus on efforts to improve the environment under which they were kept and campaigns to include the disease within a programme of eradication.

That these high levels of bovine tuberculosis in cattle generated growing concern is hardly surprising. In a depressed agricultural market, the disease represented a threat to the financial health of an already precarious agricultural industry. Estimates in 1908 placed the annual loss from bovine tuberculosis to farmers at £1 million. As one writer noted, 'from an economic point of view, tuberculosis constitutes a mortgage of Shylock variety upon a large percentage' of cattle.[14] The human cost, however, was far higher. For the *Medical Press and Circular* 'there was no disease which at present cost such a serious loss to the community or diminishes the food supply of the country so much as tuberculosis'.[15] The prevalence of the disease in cattle was correlated with levels of non-pulmonary tuberculosis in the human population. Under these circumstances, bovine tuberculosis was increasingly perceived as a serious public health problem.

A meat eating nation

Tuberculosis as the main cause of death in the nineteenth century generated considerable anxiety, especially following the identification of the tubercle bacillus, but alarm about bovine tuberculosis was more than just a component of the crusade against consumption, or anxieties about child health and national degeneration. A correlation existed between an increase in meat consumption and the emergence of debate about the boundaries between animal and human disease and the problem of diseased meat. However, the relationship between this concern, public awareness and falling meat prices is unclear.

[12] August Lydtin, George Fleming and M. van Hertsen, *The Influence of Heredity and Contagion on the Propagation of Tuberculosis and the Prevention of Injurious Effects from Consumption of the Flesh and Milk of Tuberculous Animals* (London, 1883), p. 38.
[13] *Lancet* ii (1890), p. 832; *BMJ* i (1884), p. 754; *Public Health* v (1892/3), p. 201.
[14] *Bibby's Book on Milk*. Section IV. *Bovine Tuberculosis: Cause, Cure and Eradication* (Liverpool, [1911]), p. 185.
[15] *MPC*, 22 Feb. 1888, p. 187.

Although there are methodological problems in determining how much individual items contributed to overall diets, with evidence of consumption often coming from incomplete or ambiguous sources that make it hard to draw firm conclusions, historians have argued that the standard of living rose from the mid nineteenth century.[16] In response, there was a shift in the national diet from patterns established in the eighteenth century that had favoured more grain-based foods. Changes in agricultural and industrial production linked to advancements in refrigeration and improvements in animal breeding and husbandry saw eating habits alter with a rise in household incomes and an increase in food imports. The price of staples and grain fell to produce a relative rise in urban real wages that allowed all sections of society to consume a greater volume and variety of food.[17] One result of this rise in the standard of living was an increase in meat consumption as the cost of food fell and beef prices declined. Estimates in 1861 placed meat consumption in Britain at 59.5 lb per year, one of the highest in Europe. By the end of the nineteenth century, meat had become the largest single item of food expenditure after bread. Studies for the Royal Statistical Society in 1903 demonstrated that meat consumption had risen to 107 lb for the 'wage earning class'; nearly twice this amount for the middle class.[18] As George Newman noted, by 1905 'the English working-man is, relatively speaking, a large meat eater, and, as a rule, the more wages he gets the greater the quantity of meat he eats.'[19]

Despite this general increase, the amount and quality of meat included in urban diets varied greatly. Not all benefited from the rise in meat consumption in the same way. Patterns of consumption varied geographically – for example, more meat was eaten in towns than in rural areas – and by class. Middle-class diets were rich in meat, especially as vegetables were not regarded as important.[20] Although working-class diets, which were essentially elastic and related to employment, were regularly criticised for their nutritional value and uneconomical nature largely because they were believed to rely too heavily on animal protein, for the poor meat was bought when finances permitted, often in small quantities. Many were aware of the high price of meat and the limitations this placed on purchase. For the poorest income groups, most of

[16] For a succinct summary of the debate see Charles H. Feinstein, 'Pessimism Perpetuated: Real Wages and the Standard of Living in Britain during and after the Industrial Revolution', *JEcH* xcviii (1998), pp. 625–58.

[17] Kevin O'Rourke, 'The European Grain Invasion, 1870–1913', *JEcH* lvii (1997), pp. 775–801.

[18] 'Consumption of Meat', *BMJ* i (1861), p. 159; Royal Statistical Society, *The Production and Consumption of Meat* (London, 1903), p. 15.

[19] George Newman, 'The Administrative Control of Food', *Public Health* xix (1905/6), pp. 73–4.

[20] See Michael Nelson, 'Social-class Trends in British Diets, 1860–1980', in *Food, Diet and Economic Change Past and Present*, ed. Catherine Geissler and Derek J. Oddy (Leicester, 1993), p. 103.

the meat that was purchased went into the Sunday dinner with the principal share 'bought for the men'.[21]

Although meat remained an expensive food item for the poor, the absence of meat in working-class diets came to be seen as a form of deprivation. As one study in 1915 suggested, 'a workman would sacrifice part of defined necessaries in favour of a meat diet'.[22] Doctors extolled its virtues and argued that meat 'exceed[s] all other foods in nutritional power'.[23] However, as Ross has illustrated, meat was more than a source of nutrition. Despite its high relative cost meat acquired a powerful value. It helped define status and gender and reflected notions of respectability at a time when large sections of the poor were caught up in 'the daily miracle of the loaves and fishes'.[24] Meat became an item of conspicuous consumption, a way of demonstrating financial stability. It had an important cultural significance in the national diet; every attempt was made to include it in meals, or at least to give meals a meat flavour. However, it was beef that 'stands out prominently in the eyes of all Englishmen as the most nourishing, most enjoyable, and best of its class'.[25] And it was around beef that fears about meat and the transmission of disease from animals to humans coalesced.

Selling diseased meat

Whereas historians have tended to examine the quantitative evidence of how much meat was consumed, or seen the amount of meat included in the weekly diet as a window on the poverty of the manual workforce, a neglected question is the quality of meat eaten. Although much of the evidence relies on the biased views of social commentators and sanitarians who wanted to clean up the meat trade, evidence does suggest that even with more meat being consumed, much remained of dubious quality. The high relative cost of meat in comparison to staples ensured that the poor were easily tempted into purchasing anything that looked capable of being converted into a meal. Many were reliant on bargain cuts or what was left at the end of the day. Edgar Wallace remembered poor families along the Old Kent Road in the 1890s relying on 'those "odds and ends" of meat, the by product of the butchering business' for their meat.[26] Others lamented that 'our kind of wages' ensured

[21] *Inter-Departmental Committee on Physical Deterioration*, PP (1904) xxx, p. 11.
[22] A. L. Bowley and A. Burnett-Hurst, *Livelihood and Poverty* (London, 1915), p. 80.
[23] Jonathan Pereira, A *Treatise on Food and Diet* (London, 1843), p. 435.
[24] Ellen Ross, *Love and Toil: Motherhood in Outcast London, 1870–1918* (New York, 1993), p. 32.
[25] Philip Birch, 'The Food of the Household', in *Health Lectures for the People* (London, 1885), p. 25.
[26] Cited in Anthony S. Wohl, *Endangered Lives: Public Health in Victorian Britain* (London, 1983), p. 50.

that 'we never have butchers meat', forcing them to get their meat from other sources.[27] The high relative cost of meat, low wages and long hours, especially for women in mill towns, forced many to rely on convenience foods for their meat, especially meat pies, 'various cold cooked meats' and sausages.[28] Such convenience foods were cheap because they often contained the poorest quality meat. However, it was not just a question of cheap cuts, '"odds and ends"' and convenience foods. After 1850, alarm grew about the large quantities of meat 'only fit to be buried' that were being sold.[29] Much of this was felt to come from diseased livestock; after 1870 predominantly from tuberculous cattle.

Although there is little quantitative evidence to suggest how much diseased meat was sold, records from the medieval period onwards show that sick animals regularly entered the food chain.[30] As the price of meat rose after 1850 and the market for cheap meat increased, the quality fell at the same time as increased competition and demand saw a shift in the nature of the retail meat trade to favour the growth of more small-scale butchers. Falling meat prices after 1870 may well have encouraged these butchers and traders to be less scrupulous about the meat they sold. Regular outbreaks of epizootics in the mid nineteenth century – notably rinderpest, pleuro-pneumonia, foot-and-mouth, and anthrax – flooded the market with meat from diseased livestock. It was more profitable to sell diseased cattle for food than seek veterinary attention for affected animals. Farmers, dairymen and stockowners sold diseased livestock to butchers when they started to show any sign of disease 'in order to hold their ground' in an environment in which a waste-not-want-not attitude prevailed.[31] A shadowy business in diseased meat steadily grew as a result of these factors 'which offered farmers a far greater return for diseased cattle than the paltry sum that they would receive for the dead animal's hide'.[32] According to the surgeon and former veterinary student Joseph Sampson Gamgee, 20 per cent of all meat sold in the 1850s came from diseased animals. This was a much higher percentage than earlier reports had assumed.[33]

[27] Cited in Betty McNamee, 'Trends in Meat Consumption', in *Our Changing Fare: Two Hundred Years of British Food Habits*, ed. T. C. Barker, J. C. McKenzie and John Yudkin (London, 1966), p. 77.

[28] Elizabeth Roberts, *A Woman's Place: An Oral History of Working-Class Women, 1890–1940* (Oxford, 1984), p. 159.

[29] Joseph Sampson Gamgee, *Cattle Plague and Diseased Meat in their Relations with the Public Health . . . A second letter to Sir G Grey* (London, 1857), pp. 5–6.

[30] John D. Blaisdell, 'To the pillory for putrid poultry: Meat Hygiene and the Medieval London Butchers, Poulterers and Fishmongers' Companies', *Veterinary History* ix (1997), pp. 114–24.

[31] John Gamgee, *The Diseases of Animals in Relation to Public Health and Prosperity* (Edinburgh, 1863), p. 11.

[32] Joanna Swabe, *Animals, Disease and Human Society* (London, 1999), p. 97.

[33] Joseph Sampson Gamgee, *Cattle Plague and Diseased Meat in their Relations with the Public Health and with the interests of Agriculture. A letter to . . . Sir G Grey* (London, 1857), p. 2.

Contemporaries by the mid nineteenth century were all too aware that 'bad meat unfit for food' made up a substantial proportion of the meat sold.[34]

London was believed to be at the centre of this trade. The newly formed Metropolitan Association of Medical Officers of Health (MAMOH) was convinced that 'a large quantity of unwholesome and diseased meat is daily sold' in London.[35] The trade in diseased meat was made easier by inadequate inspection (see chapter 5) and by conditions in meat markets and slaughter-houses, 'places disgraceful to any large city'. Individual markets were considered to be overwhelmed with diseased meat. When the veterinarian and reformer John Gamgee, brother of Joseph Sampson, investigated the Newgate meat market in London in 1857, he explained that

> the vast number of its dark little shops, or rather holes offers great facility to the hiding [of] bad meat, which in the day is perfectly visible and when brought out under a gas illumination on Saturday night does not show its true colours, and finds purchaser in the poor and hard working population.[36]

He noted that no attempt was made to disguise the trade. Gamgee was not alone in expressing alarm. An American physician visiting London in 1862 explained how

> in my visits to the various markets in London, and especially those narrow, dark streets, where meat is sold to the poorest classes, I have been greatly surprised to see the great quantities of diseased meat allowed to be sold. I have seen, within the sound of Bow bells, more diseased meat offered for sale during the last two months, than I ever saw in New York during my whole life.[37]

Thirty years later, Sedgwick Saunders, the MOH for the City of London, could still claim that 'for many years past it has been notorious that enormous quantities of unsound and diseased meat have been sold with impunity' in London.[38]

However, the practice of selling diseased meat was not limited to London. Contemporaries pointed to how the trade in diseased meat was 'so common a practice that it was not looked upon as a crime'.[39] In the Newton Heath district of Manchester, for example, three of the sixteen slaughterhouses in the area

[34] CLRO: Sanitary committee, 18 Oct. 1870.

[35] MTG, 30 Aug. 1856, p. 218.

[36] Gamgee, *Cattle Plague and Diseased Meat . . . A letter to . . . Sir G Grey*, p. 39.

[37] *BMJ* ii (1862), p. 287.

[38] CLRO: Sanitary committee, 12 Mar. 1895.

[39] 'Is Selling Diseased Meat Immoral?', *Sanitary Record*, 30 Oct. 1875, p. 235; *BMJ* i (1863), p. 17.

were believed to specialise in the killing and preparation of diseased carcasses.[40] Similar practices were reported in Oldham. Some butchers made their living from selling unsound and diseased meat. It was believed that the 'custom is more or less prevalent in all large towns', but it was not just the disreputable who sold diseased meat.[41] Disease was considered so rife in cattle in the 1860s that it was felt that 'the salesman cannot live unless he gets a price for diseased meat which falls into his hands'. This ensured that an 'enormous quantity' of diseased meat was 'daily thrust upon our markets'.[42] Although such claims were no doubt a piece of journalistic exaggeration, the Privy Council estimated in 1862 that one-fifth of butchers sold meat from animals that were 'considerably diseased'. This is not to suggest that the whole meat industry from the farmer to the butcher was involved in knowingly selling diseased meat. However, with good profits to be made, a poor system of detection, and relatively low penalties if caught, it was acknowledged that the trade in diseased meat was ubiquitous.[43]

By the 1880s, concerns about diseased meat had come to focus on the question of bovine tuberculosis. Contemporaries could point to the substantial amounts of tuberculous meat that was 'almost daily being carried into the large towns'.[44] Rural districts were gradually identified as reservoirs of infection. In Glasgow, for example, it was felt that two-thirds of the 'suspicious or diseased meat' dealt with came 'from the country'.[45] The problem was perceived as grim, especially in London where alarmists claimed that 80 per cent of the meat sold exhibited the telltale tubercles. Despite the figure being widely dismissed as ridiculous, Gibson Bott, chairman of the London County Council's public control committee, was all too aware that 'there is reason to fear' that tubercular meat 'finds its way into the market'.[46] Other contemporaries noted similar problems. Francis Vacher, MOH for Birkenhead and an expert on meat inspection, explained that the disease was so common in the town that it was 'quite impracticable' to exclude it from the meat market; in Yorkshire, James Higgins, the meat inspector, reported a 'flourishing business' in tuberculous meat in Leeds and the surrounding towns. Comparable practices were reported

[40] *Report to the General Board of Health on a preliminary inquiry into the Sewerage, Drainage, and Supply of Water, and the sanitary condition of the inhabitants of Newton Heath* (London, 1852), p. 52.

[41] Headlam Greenhow, *Report on Murrain in Horned Cattle, the Public Sale of Diseased Animals, and the Effects of Consumption of their Flesh on Human Health* (London, 1857), p. 46; *Select Committee Appointed to Consider the Operation of the Acts for the Prevention of Infectious Disease in Cattle*, PP (1850) xxiii, pp. 21–2.

[42] *BMJ* i (1863), p. 17.

[43] *Fifth Report of the Medical Officer of Privy Council for 1862* (London, 1863), p. 22.

[44] Lydtin, Fleming and van Hertsen, *Heredity and Contagion*, p. 168.

[45] GCA: Special subcommittee on diseased meat, 14 Feb. 1882, MP9.26.

[46] Henry Behrend, *Cattle Tuberculosis and Tuberculous Meat* (London, 1893), p. 12; LMA: LCC minutes, 3 Feb. 1893, p. 137.

elsewhere.[47] What the public would not buy was frequently sold to pie and sausage manufacturers. Rising prices and low fines if apprehended ensured that the trade remained a profitable one.

Although it was feared that the middle classes and their cooks could be duped into buying unwholesome or diseased meat, there was a clear market for such meat among the urban poor and working class. Given the cultural value assigned to meat, the poor, 'even the thrifty poor', as one journal noted, were forced to keep down the cost of their food by 'purchasing inferior, often unfit' meat.[48] Butchers killed animals of poor quality to produce 'poor meat to supply poor people at a poor price', arguing that if they did not 'many people would hardly ever taste flesh meat at all'.[49] Most diseased and tuberculous meat was, as Gibson Bott explained, therefore 'sold largely amongst the poor' either directly or in the form of sausages and other convenience foods.[50] In poor neighbourhoods it was a matter of notoriety that the meat exposed for sale was generally diseased. Diseased and suspect meat was sold cheap, especially at Saturday night markets 'when necessitous people' bought up such meat 'at reduced prices from salesmen's shop[s]' for their Sunday dinner.[51] Diseased meat proved 'irresistible bait to hungry mouths'.[52] That the poor were eating diseased meat aroused considerable alarm, but the problem was felt to go beyond the poor. As one doctor explained in 1856, 'Even the higher classes were not exempt' from the sale of diseased meat.[53]

Man, animals and meat

Contemporaries were all too aware in the 1850s that 'it is beyond question that the people are already very largely consuming diseased meat as food'.[54] By 1889, the connection between diseased meat and health had become an issue of national importance as part of a wider discussion about food quality and safety and concerns about poverty and health. However, a sense that diseased meat was prejudicial to health was present before the nineteenth century. Awareness about the dangers of eating 'unwholesome' or diseased meat

[47] *Royal Commission Appointed to Inquire into the Administrative Proceedings for Controlling the Danger to Man through the use as Food of the Meat and Milk of Tuberculosis Animals*, minutes of evidence (London, 1898), p. 21; Gerald Leighton and Loudon Douglas, *The Meat Industry and Meat Inspection* 5 vols (London, 1950), vol. iv, p. 1185; James Higgins, 'On Diseased Meat', *EVR* v (1863), pp. 670–1; 'Diseased Meat', *Sanitary Record*, 3 Apr. 1875, p. 233.
[48] *The Times*, 6 Apr. 1863, p. 6.
[49] 'Diseased Meat Cases', *Sanitary Record*, 16 Sept. 1889, p. 143.
[50] LMA: LCC minutes, 3 Feb. 1893, p. 137.
[51] CLRO: Sanitary committee, 24 Mar. 1896.
[52] *Lancet* ii (1871), p. 746.
[53] *The Times*, 10 Mar. 1856, p. 11.
[54] Gamgee, *Cattle Plague and Diseased Meat . . . A letter to . . . Sir G Grey*, p. 2.

can be detected in ideas about meat hygiene in the medieval period. With a shift in diets from scarcity to 'cheap and plentiful food' by the end of the fourteenth century and a growth in the amount of meat sold in urban markets the idea that meat from diseased animals could cause sickness was more clearly articulated. A connection was made between rotten and obviously diseased meat and illness. Numerous measures were therefore adopted to prevent butchers from selling 'unhealthy or corrupt flesh'.[55] Outbreaks of epizootics prompted additional measures. For the main part, however, the regulation of meat supply was left to local guilds. In medieval London, the Honourable Company of Butchers was responsible for setting standards for the slaughtering of animals. Evidence suggests that the company and its wardens made a connection between animal and human disease as outbreaks of animal plague in the mid fourteenth century had seen efforts to prevent the sale of diseased meat.[56] It was well known that 'bad meat' could 'trouble and offend the stomacke'. By the sixteenth century, links were made between sickness, corruption and 'unwholesome' or diseased meat as part of a humoral understanding of ill health.[57] However, regulation and a sense that the meat trade was honest initially deflected attention. This did not mean that prosecutions for selling 'blown or bad meat' were halted; nor that measures to control the meat trade were abandoned, but that little concerted effort was made to police the meat trade.[58]

The eighteenth century saw renewed interest in the relationship between diseases in animals and in humans. The possibility that some diseases could be spread from animals to humans was recognised at a popular and medical level. This was clear in folk traditions that cowpox could be contracted by dairymen. Outbreaks of epizootics became increasingly well documented and experimental studies in Europe on the contagious diseases of animals focused on the idea of transmission.[59] It was a simple step to assume that meat from

[55] Christopher Dyer, 'Did the Peasants really Starve in Medieval England?', in *Food and Eating in Medieval Europe*, ed. Martha Carlin and Joel Rosenthal (London, 1998), p. 70; Martha Carlin, 'Fast Food and Urban Living Standards in Medieval England', *ibid.*, pp. 27–51.

[56] Blaisdell, 'To the pillory for putrid poultry', pp. 114, 117.

[57] *The philosophie, commonlie called, the morals vvritten by the learned philosopher Plutarch of Chæronea. Translated out of Greeke into English, and conferred with the Latine translations and the French, by Philemon Holland of Coventrie, Doctor in Physicke* (1663), p. 226; see William Kemp, *A brief treatise of the nature, causes, signes, preservation from, and cure of the pestilence collected by William Kemp* (London, 1665), p. 14; Obadiah Walker, *Of education, especially of young gentlemen in two parts, the second impression with additions* (1673), p. 199.

[58] *ORDERS Conceived and Published by the Lord Mayor and Aldermen of the City of London, concerning the Infection of the Plague* (London, 1665), p. 54.

[59] See Lise Wilkinson, 'Zoonoses and the Development of Concepts of Contagion and Infection', in *The Advancement of Veterinary Science: The Bicentenary Symposium Series*,

such animals was harmful to human health. Ideas that unwholesome meat had a 'putrid tendency', was likely to become 'putrid in the stomach', enter the blood and cause disease gained ground and were given a chemical rationale.[60] Medical writers cast the stomach as a site of disorder or suggested that the alkaline nature of meat promoted inflammation and fever as links were made between diet and health. Food became a remote cause of disease that could weaken the constitution.[61]

Although Hardy has asserted that it was not until 'the new bacteriology began to uncover the specific agents of disease in the years after 1876 that evidence began to emerge that could begin effectively to establish links between animal and human disease', this did not stop doctors in the mid nineteenth century from exploring the connection between food, animals and disease.[62] In the 1840s, chemists and physiologists began to relate the chemistry of food to animal physiology. Much of the work focused on the nutritional value of food, but Liebig's *Animal Chemistry* (1842) did introduce new ideas about the relationship between the food that was consumed and its impact on the body.[63] Although most of this work examined the properties of food, studies in 1844 by Robert Christison, professor of materia medica and therapeutics at Edinburgh University, suggested that products from diseased animals might cause poisoning. Investigations into rabies in the early nineteenth century also began to demonstrate that a disease could be experimentally transmitted from humans to animals through inoculation. Apprehension was voiced about the consequences of eating meat from diseased animals and reports multiplied about the evils of the meat trade just as growing faith was being placed in the

vol. iii, *History of the Health Professions: Parallels between Veterinary and Medical History*, ed. A. R. Michell (Oxford, 1993), pp. 76–80; Lise Wilkinson, 'Glanders: Medicine and Veterinary Medicine in Common Pursuit of a Contagious Disease', *Medical History* xxv (1981), pp. 363–84.

60 William Smith, *A Sure Guide in Sickness and Health, in the Choice of Food, and Use of Medicine* (London, 1776), p. 7.

61 John Woodward, *The State of Physic: And of Disease; With an Inquiry into the Causes of the Late Increase in Them* (London, 1718); William Buchan, *Dr Buchan's Domestic Medicine; Or a Treatise on the Prevention and Cure of Disease, by Regimen and Simple Medicine* (Newcastle, 1812); George Cheyne, *The English Malady: Or, A Treatise of Nervous Diseases of All Kinds, as Spleen, Vapours, Lowness of Spirits, Hypochondriacal, and Hysterical Distempers, etc* (London, 1733); *idem*, *An Essay on Regimen* (London, 1740). For Cheyne, see Anita Guerrini, *Obesity and Depression in the Enlightenment: The Life and Times of George Cheyne* (Norman, Oklahoma, 2000); Steven Shapin, 'Trusting George Cheyne: Scientific Expertise, Common Sense and Moral Authority in Early Eighteenth-century Dietetic Medicine', *BHM* lxxvii (2003), pp. 263–97.

62 Anne Hardy, 'Pioneers in the Victorian Provinces: Veterinarians, Public Health and the Urban Animal Economy', *Urban History* xxix (2002), p. 376.

63 Harmke Kamminga and Andrew Cunningham, 'Introduction: The Science and Culture of Nutrition, 1840–1940', in *The Science of Nutrition, 1840–1940*, ed. Harmke Kamminga and Andrew Cunningham (Amsterdam, 1995), pp. 3, 4.

value of 'cheaper and more abundant beef' in maintaining health and in the importance of animal protein in the diets of the labouring classes.[64]

Why did diseased meat start to become a source of concern in the 1850s? A growing urban population increased the need for greater numbers of livestock to be driven in, marketed and slaughtered which made the sale of diseased animals more visible. At the same time, the retail price of meat and outbreaks of contagious disease in cattle, and in particular pleuro-pneumonia, were rising. As food animals were absorbed into the workings of the urban economy and patterns of consumption, curiosity grew about the relationship between animals, meat and disease as part of an emerging critique of food safety. This anxiety about the health consequences of eating meat from diseased livestock occurred against a background of mounting alarm about food and the dangers of adulteration.[65] In debates about diseased meat, a similar stance to that adopted in discussions about adulteration was initially taken. It was suggested that the public were being cheated by the sale of diseased meat, since such poor quality meat robbed the consumer 'of his fair share of nourishment'. Growing evidence about the 'evils' of the meat trade and a sense that the poor were being forced to rely on cheap and inferior quality meat focused attention. The appointment of MOsH, and the beginnings of professional organisations among them, provided a framework within which these concerns about food, disease and meat could be debated.

From the start, concerns about the health consequences of eating diseased meat did not fit with a hygienic gospel of cleanliness and purity that characterised public health and debates on adulteration. It was more a question of poisoning than fraud. As the idea that diseased meat 'may be a specific cause of disease' gained ground throughout the 1850s and 1860s, encouraged by a series of epizootics in cattle, calls were made for a 'more extended enquiry into the laws of health governing the health and diseases of man as affected by the health and diseases of animals'.[66]

In the mid Victorian period, relatively little was known about the transmission of disease from animals to humans. Part of the reason stemmed from what Hardy has referred to as the 'opaque nature' of the transfer of animal diseases to humans. She has argued that the idea of accepting that 'the diseases of humans and animals might be causally related required a cultural and intellectual readjustment'. Before the mid nineteenth century, infection from animal diseases could either be thought of as accidental, such as via inoculation, as in the case of anthrax or rabies, or following a process that was long

[64] Mark R. Finlay, 'Quackery and Cookery: Jutus von Liebig's Extract of Meat and the Theory of Nutrition in the Victorian Age', *BHM* lxvi (1992), p. 408.

[65] For a history of adulteration in the early nineteenth century, see John Burnett, *Plenty and Want: A Social History of Diet in England from 1815 to the Present Day* (London, 1966).

[66] Cited in Richard Perren, *The Meat Trade in Britain, 1840–1914* (London, 1978), p. 52.

enough so as to produce an obscure 'pathway of causation'.[67] Nor was there a framework for study. The low status of the veterinary medicine, and the tendency for dairymen and stockowners to call in 'some nostrum vendor or other quack' to treat diseased livestock, hampered investigation.[68] Until the 1860s, veterinary medicine was far from being a profession. Although a number of veterinary schools had been established in London, Glasgow and Edinburgh, Britain was behind its European counterparts. Teaching in these institutions was predominantly practical and tended to be dominated by equine medicine, the mainstay of most veterinary practices. Little attention was therefore directed at the 'diseases of animals of the farm', especially as beasts that showed any signs of disease were quickly slaughtered and sold.[69] In addition, interest in comparative anatomy had declined following John Hunter's work in the late eighteenth century. The subject had become marginalised in medical education, and it was not until the 1880s that it started to attract renewed attention as part of a debate about the pathology of contagion and veterinarians' attempts to assert their scientific credentials.

Discussion about the threat posed by eating diseased meat was initially limited to a small group of MOsH and elite veterinarians, many of whom were trying to affirm their professional position at a local level. At the time, public health was dominated by the need to resolve the structural and sanitary problems facing many towns, and by debates about the causation of epidemic disease. However, despite the relatively small numbers of doctors who were at first disturbed by the large quantities of diseased meat sold, interest went beyond an awareness of the effects of decomposition and the role played by ptomaine poisoning.[70] It came to focus on the assumption that diseased meat should be dangerous. Theories were put forward that linked specific animal diseases to meat. These views were combined with a sense that 'the same causes which promote sickness in the lower animals may with just reason be suspected of favouring disease amongst ourselves'.[71] Here the work of John Gamgee was central in promoting interest.

John Gamgee had initially trained as a doctor before switching to veterinary medicine. After studying in Europe, where the standard of veterinary education was considerably higher than in London, he returned to England and set about reforming veterinary training, establishing the New Edinburgh Veterinary College in 1857 and the *Edinburgh Veterinary Journal* in the following year. His knowledge of disease pathology and aetiology and his clinical experience

[67] Anne Hardy, 'Animals, Disease and Man', *Perspectives in Biology and Medicine* xlvi (2003), p. 203.

[68] Greenhow, *Murrain*, p. 12.

[69] See Iain Pattison, *The British Veterinary Profession, 1791–1948* (London, 1984); *Veterinarian* xxxv (1862), p. 276.

[70] For a discussion of food poisoning see Anne Hardy, 'Food, Hygiene and the Laboratory: A Brief History of Food Poisoning in Britain', *SHM* xii (1999), pp. 293–311.

[71] *British and Foreign Medico-Chirurgical Review* xxi (1858), pp. 88, 87.

were superior to those of his colleagues.[72] Gamgee was an outsider who challenged veterinary orthodoxy and rejected notions of spontaneous generation commonly accepted by doctors and veterinarians. Instead, he adopted a contagionist model of infection. He was also convinced that certain cattle diseases, notably pleuro-pneumonia, could cross the species barrier through the medium of diseased meat.[73] It was in Edinburgh, following a series of epizootics, that Gamgee started campaigning for a reform of the meat trade, which won him influential backing for his veterinary college and attracted national attention. He was appalled by what he found: Gamgee witnessed how organs and flesh from diseased cattle were regularly combined with other meat and used in sausages before being sold to an unsuspecting public. Through a series of letters to the *Scotsman* and the *Glasgow Herald*, he railed against these abuses, calling for moves to stamp out the sale of diseased meat.[74]

In these campaigns, John was joined by his brother Joseph Sampson, assistant surgeon at the Royal Free Hospital and president of the Medical Society of University College London. In open letters to the press and the home secretary, they highlighted the evils of the meat trade and the prevalence of diseased meat in urban markets. Both were convinced that 'the public health is materially affected by the wholesale slaughter of diseased animals as human food.' As evidence cases were cited whereby butchers had fallen ill after slaughtering diseased cows and a range of cattle diseases that included anthrax, 'splenic apoplexy' and pleuro-pneumonia were presented as injurious to human health. The Gamgee brothers argued that doctors need only look to the role parasites played in the spread of diseases like trichinous for confirmation that meat from diseased livestock was hazardous to health.[75]

The Gamgees were often ridiculed and accused of arrogance, but they were not alone. Leading public health campaigners in London like Benjamin Ward Richardson were also starting to draw attention to the dangers of diseased meat.[76] Provisions under the 1851 City of London Sewers Act and the Metropolitan Market Act of the same year had seen the appointment of an inspector of meat and slaughterhouses and of markets. In addition, the appointment of MOsH in London created a network of public health doctors whose duties included the inspection and regulation of slaughterhouses and markets. These

[72] See John R. Fisher, 'Professor Gamgee and the Farmers', *Veterinary History* i (1979–81), pp. 47–63; Sherwin A. Hall, 'John Gamgee and the Edinburgh New Veterinary College', *Veterinary Record* lxxvii (1965), pp. 1237–41.

[73] Gamgee, *Diseases of Animals*, pp. 6–7.

[74] See John Gamgee, *Diseased Meat sold in Edinburgh, and Meat Inspection, in connection with the Public Health, and with the Interests of Agriculture. A Letter to the . . . Lord Provost of Edinburgh* (Edinburgh, 1857).

[75] John Gamgee, 'A Case Illustrating the Dangers of Slaughtering Diseased Cattle', *Lancet* i (1864), p. 182.

[76] See 'Sale of Bad Meat in London', *Journal of Public Health and Sanitary Review* ii (1856), p. 311.

posts helped direct attention to the widespread sale of diseased meat in London as metropolitan sanitary officials started to worry about what meat should or could be condemned. London's officers as a group were 'energetic proponents of preventive medicine' and their interest in diseased meat should be seen as part of their wider concerns about sanitary science and the need to prevent the spread of disease.[77]

In the light of these concerns, the MAMOH undertook as one of its first duties to 'inquire into the quality of animal food sold in our thoroughfares'. Already anxious about adulteration, the Association's investigation reflected metropolitan medical officers' interests in questions of pure foods and nuisances.[78] The inquiry was designed to determine the ill effects of eating diseased meat and recommended guidelines on how it might best be identified. Concern was essentially practical. Despite the report being hastily put together and carrying little scientific weight, it did highlight the threat of diseased meat to health. In addition, it sparked discussion about the spread of disease from animals to humans.[79]

Outbreaks of pleuro-pneumonia and anthrax further heightened anxiety about the dangers of diseased meat especially following reports from John Burdon Sanderson, MOH for Paddington, that diseased cattle were regularly sent to 'disreputable' slaughterhouses and 'from thence the meat finds its way into the markets'.[80] Prosecutions for the sale of 'unwholesome' meat started to be reported in the medical press and the number of court cases for the sale of diseased meat rose throughout the 1850s and 1860s as interest in the amount of diseased meat sold grew.[81] Medical journals started to call for legislation against this apparent evil of the meat trade and further investigation was demanded. In 1862, following an address by Gamgee on diseased animals and unwholesome meat, the MAMOH reaffirmed its belief that meat from animals showing signs of acute inflammation or wasting should be considered 'unfit for human consumption'. The *BMJ* added its support.[82] However, doctors remained divided as to meat's role in the causation of disease, with similar uncertainty expressed in Europe and the United States. Whereas metropolitan MOsH were clear about the injurious effects of diseased meat on health, sceptics rightly argued that there was little direct evidence of a connection

[77] Anne Hardy, 'Public Health and Experts: The London Medical Officers of Health', in *Government and Expertise: Specialists, Administrators and Professionals, 1860–1919*, ed. Roy MacLeod (Cambridge, 1988), p. 130.
[78] Wellcome: MAMOH general meeting, 9 July 1856, SA/SMO/B.1/1.
[79] 'Report on Unwholesome Meat', *BMJ* iv (1856), p. 751.
[80] *Ibid.*, p. 50.
[81] *Lancet* i (1861), p. 378; *Report of the Health of Liverpool during the Year 1868* (Liverpool, 1869), p. 53 and *ibid.* (Liverpool, 1875), p. 46 for the increase in meat seized at a local level.
[82] Wellcome: MAMOH general quarterly meeting, 16 Apr. 1862, SA/SMO/B.1/1; *BMJ* i (1863), p. 17.

between diseased meat and health. They worried that moves to condemn meat from diseased livestock would force the price of meat up and hit the poor. These discrepancies in part emerged because different forms of disease were being referred to. When pressed, however, even those who felt that 'as a general principle' diseased meat was dangerous could point to few cases to support their views, relying instead on impressionistic evidence and a sense that diseased meat *should* be hazardous.[83] Prosecutions for the sale of diseased meat highlighted these differences in the medical community as expert witnesses disagreed as to whether the meat in question was diseased.[84]

Growing apprehension in the medical and veterinary press about the possible harmful effects of diseased meat encouraged the Privy Council to appoint an inquiry under Gamgee to investigate the livestock trade. The Privy Council under John Simon, Britain's senior public health official from 1855 to 1876, was committed to a programme of investigation, and commissioned pathological and epidemiological studies that it hoped would provide persuasive scientific research to support public health efforts. Simon was keen to promote studies into the 'intimate pathology of contagion' and was also concerned about the probable health consequences of diseases of food animals for humans, although he did not share Gamgee's alarmist views.[85] Gamgee also had supporters in the Commons. They repeated his fears about diseased cattle and the threat to human health and encouraged investigation.[86] Most of the report focused on the extent to which cattle, pigs and sheep were diseased. Gamgee was primarily interested in asserting a contagious model to explain how epizootics were spread, implicating the role played by imported livestock in transmitting disease. The conditions he recounted were appalling, but he was unable (or unwilling) to come to any firm conclusions on the exact danger represented by meat from diseased livestock. Simon therefore called for further investigation.[87]

The outbreak of cattle plague (rinderpest) in 1865 focused attention among medical men on cattle diseases and microbes.[88] The speed at which the disease spread and the impact on trade were alarming: by 1866, the disease had claimed 10 per cent of the national herd, with the estimated loss between £8 and £14 million. It was, as Fisher notes, the 'most dramatic episode in nineteenth

[83] John Gamgee, 'Cattle Diseases in Relation to Supply of Meat and Milk', *Fifth Report of the Medical Officer of the Privy Council for 1862*, appendix iv (London, 1863), p. 287.
[84] See *Lancet* ii (1864), pp. 697–8.
[85] E. Seaton, *Public Health Reports of John Simon* 2 vols (London, 1887), vol. ii, pp. 75–90.
[86] *Hansard* clxxii, 2 July 1863, cols 43–4.
[87] *Fifth Report of the Medical Officer of the Privy Council for 1862* (London, 1863), pp. 206–7, 25.
[88] See Sherwin A. Hall, 'The Cattle Plague of 1865', *Medical History* vi (1962), pp. 45–58; Michael Worboys, *Spreading Germs: Disease Theories and Medical Practice in Britain, 1865–1900* (Cambridge, 2000), pp. 49–56.

century British agriculture'.[89] Rinderpest not only highlighted the poor position of veterinarians and their ignorance of cattle disease, but it also made animal disease a concern of central government, helping to launch investigations into germs. Bills had been presented to parliament to stop diseased meat from being imported and to prevent the 'exposure' of diseased cattle for sale in the early 1860s, but no action had been taken.[90] The alarming spread of rinderpest changed the situation. Farmers lobbied for greater restrictions on imports of potentially diseased livestock. Anxieties were voiced in the medical press about the impact on human health, with concern shaped by the ongoing debate on diseased meat. For the *Lancet*, 'it would seem in the highest degree improbable that the flesh of animals infected by so virulent a disease as steppe murrain could be eaten with impunity'.[91] In London, a special slaughterhouse was set aside in the Metropolitan Cattle Market for the reception of 'beasts suffering from cattle plague' from which the meat was immediately sent to the boilers for 'disinfection'. Islington also appointed a veterinary surgeon 'to guard the good people of Islington against the chance of eating sirloins from off the carcasses of the beasts at that time dying by hundreds in their cowsheds'.[92] Growing public anxiety about the safety of meat from infected cattle was reported, and it was suggested that the public should be 'protected from the sale of unwholesome meat'.[93] In response, a royal commission was appointed and several researchers were hired to investigate the aetiology of the disease. Much of the work into the nature of contagion and the pathology of rinderpest was pioneering and helped to define experimental pathology. From the start, the researchers examined the connection between rinderpest and human diseases, most commonly smallpox.[94] In the process, they stimulated further debate and work on the relationship between diseases in animals and in humans.

Cows and disease

Although doctors were quick to dismiss the possibility that meat from cattle with rinderpest could spread the disease to humans, a view confirmed by the royal commission, the cattle plague did draw attention to the potential threat posed by meat from diseased livestock and helped assert a contagious model. At the same time, restrictions on the movement of cattle following the

[89] *Saturday Review* xxi (1866), pp. 46–7, 78–80; John R. Fisher, 'Cattle Plagues Past and Present: The Mystery of Mad Cow Disease', *Journal of Contemporary History* xxxiii (1998), p. 215.
[90] *Lancet* i (1864), p. 278.
[91] *Lancet* ii (1865), p. 207.
[92] 'Cattle Plague and Diseased Meat', MTG, 16 Dec. 1865, p. 657.
[93] 'Cattle Plague', *ibid.*, 5 Aug. 1865, p. 160.
[94] Worboys, *Spreading Germs*, p. 44; *Royal Commission on the Cattle Plague*, PP (1865) xxii.

epidemic saw the dead meat trade expand and with it apprehension about the supply of diseased meat. Further anecdotal evidence was provided about the dangers of diseased meat, and cases of food poisoning were confused with evidence that meat from diseased livestock resulted in illness.[95] Pressure was applied by the MAMOH to make it easier to convict those found selling diseased meat as MOsH voiced ongoing fears that disease could be 'taken into the stomach with the food or drink'.[96] Further studies were commissioned by the Privy Council as part of a series of investigations into the 'pathology and pathogenesis of disease'. Interest initially focused on animal pathology, trichina spiralis, and the disease that resulted from eating meat containing the tapeworm, which was believed to be capable of producing a typhoid-like fever.[97] In trichinous there was a clear link between meat and disease and this helped fuel the view that meat from diseased livestock was dangerous.

However, too many questions remained for an active campaign against the sale of diseased meat. As Hardy has noted, 'social and gastronomic memory and experience favoured the assumption of a benign and harmonious interchange with the natural world' and with food.[98] Anxieties continued to be voiced about the circumstantial nature of the evidence, and about the implications of banning all diseased meat, as it was predicted that 'an enormous amount of animal food' would have to be destroyed.[99] Instinct suggested that diseased meat was abhorrent and harmful to health, but doctors sought precise, definite facts to back up their prejudices. They valued the type of empirical observational evidence that was missing from discussions of diseased meat.

It was only with new mechanisms to prevent the sale of meat 'unfit for human consumption' under the 1875 Public Health Act and further pathological studies that emphasised the dangers of meat, particularly meat from tuberculous livestock, that concerted attempts were made to regulate urban meat markets. Throughout the 1870s, an increasing number of MOsH began to warn of the amount of diseased meat exposed for sale and the number of prosecutions for the sale of diseased meat rose dramatically. In the process, concern came to focus on the question of tuberculosis, so that by 1880 diseased meat had become synonymous with tuberculous meat. Tuberculosis broke the 'conceptual barrier' that distinguished the diseases of animals from those of humans.[100]

[95] See 'Diseased and Unwholesome Meat', MTG, 5 Jan. 1867, p. 21.

[96] *Report of the Sanitary Condition of Leicester in 1870* (Leicester, 1871), p. 6.

[97] John J. W. Thudichum, 'Parasitic Diseases of Quadrupeds used for Foods', *Seventh Report of the Medical Officer of the Privy Council for 1864*, appendix vii (London, 1865), pp. 303–467.

[98] Hardy, 'Animals, Disease and Man', p. 201.

[99] *BMJ* ii (1863), pp. 504–6.

[100] Hardy, 'Animals, Disease and Man', p. 203.

3

Coughing cows

Bovine tuberculosis and contagion

Looking back, Edward Ballard in a paper to the Hygienic Congress in 1891 noted that 'much of the etiological research of the last 20 years has had reference to the production of disease in man through the agency of food'.[1] By the time Ballard was writing, bovine tuberculosis had come to stand at the centre of debates about food safety and had emerged as the model zoonotic disease. However, thirty years earlier the situation had not been so clear. Interest in zoonoses focused on glanders, rabies and pleuro-pneumonia, and the relationship between cattle diseases and food remained imprecise. Studies only began to focus on the association between tuberculosis in cattle and the disease in humans in the 1860s as other cattle diseases were dismissed as posing little danger to health. The effects of pleuro-pneumonia and the devastation caused by rinderpest had initially absorbed attention, but at the same time paved the way for investigations into bovine tuberculosis.

That a link was established between bovine and human tuberculosis in the 1860s was hardly surprising, however. Pathologists started to look more seriously to contagious animal diseases for answers about the way diseases spread and the 'pathology of contagion'. Research into animal diseases contributed to a broader understanding of the aetiology and patho-physiology of many diseases and was 'a source of key moments in the history of germ ideas and practices'.[2] Interest in bovine and human tuberculosis was part of this broader inquiry and drew on the growing discipline of comparative medicine to make connections between the disease in cattle and the disease in humans. Although European pathologists led the way, in Britain veterinarians were the first to draw conclusions from these studies and assert the idea that bovine tuberculosis was contagious. Despite growing alarm about the harmful effects of eating diseased meat, doctors were more sceptical and were slower to see bovine tuberculosis as a threat to human health. The different interpretations taken reflected the contrasting conceptions of disease adopted by veterinarians and doctors and demonstrated the uneven spread of ideas about disease between countries and between practitioners.

[1] Edward Ballard, 'On Meat Infection', *Veterinary Record*, 29 Aug. 1891, p. 118.

[2] Michael Worboys, *Spreading Germs: Disease Theories and Medical Practice in Britain, 1865–1900* (Cambridge, 2000), p. 42.

Although confusion continued to exist, reinforced by uncertainties about the nature of tuberculosis, by 1882 the weight of pathological investigation asserted a common identity between human and bovine tuberculosis and its contagious properties. Robert Koch's announcement to the Berlin Physiological Society in 1882 that he had identified the agent responsible for tuberculosis served to recast the debate in bacteriological terms. However, Koch's claims that bovine tuberculosis was identical to human tuberculosis and thus transmissible to humans merely confirmed what many already suspected. What Koch did do was settle the question of the relationship between tuberculosis in animals and the disease in humans and its status as a contagious disease. Pathologists and others interested in germs reinforced Koch's work, and in the process constructed a consensus that bovine tuberculosis was a threat to public health that had to be tackled.

Framing bovine tuberculosis

The symptoms of pulmonary tuberculosis were well known to Victorian medical practitioners. Yet alongside this familiarity, there was also uncertainty about the nature and cause of tuberculosis. Interpretation of the morbid appearances of tuberculosis and of their relation to each other depended upon an ever-shifting theory of the disease. Whereas some doctors in the early nineteenth century speculated that it might be transmissible, others linked tuberculosis to a range of causes from heredity, poor nutrition and environmental factors to a weak physical constitution. Some clinicians, following what became known as the 'German school', saw tuberculosis as one possible outcome of inflammation and a common degenerative process rather than a specific infectious disease. However, it was the findings of Rene Laënnec that dominated interpretations until the 1860s. Working within an environment in which localist pathologies were gaining ascendancy, he defined consumption as one of a number of tubercular afflictions characterised by the presence of small swellings or nodules and saw the presence of the characteristic tubercles in other diseased organs as suggestive that there were different forms of the same disease.[3] Yet, despite these shifting interpretations, most could agree that tuberculosis was the result of predisposing and hereditary causes, ensuring that consumptives and their families carried the stigma of an inherited 'taint'. In cattle, the same framework of tubercular diathesis was adopted. Tuberculous cattle were seen as the offspring of inferior livestock, which had been further weakened by poor conditions, bad hay or dirty water.[4]

[3] R. Maulitz, *Morbid Appearances: The Anatomy of Pathology in the Early Nineteenth Century* (Cambridge, 1987), pp. 71–80, 94–100.
[4] See John Sherer, *Rural Life* (London, 1868).

From the 1860s, these ideas about bovine and pulmonary tuberculosis started to be challenged as part of attempts to examine methods of contagion. Here the inoculation experiments of Jean-Antoine Villemin, a physician in the French cavalry, proved important in encouraging a connection to be made between bovine and human tuberculosis and in raising questions about its contagious nature. France had pioneered the establishment of veterinary schools in the eighteenth century, partly to encourage the study of cattle plagues. These schools and further investigations into the aetiology of contagious animal diseases, many of which had important economic implications, provided a framework for the development of comparative medicine and study into bovine tuberculosis.[5] Laënnec had already affirmed the unitary nature of tuberculosis, whilst the first successes of Louis Pasteur and early work on glanders, rabies and anthrax had stimulated interest in the comparative pathology of infectious disease and pointed to the power of the laboratory. In light of this work, a number of French veterinarians and clinicians became interested in aspects of pathogenesis and animal diseases as part of broader investigations into the nature of disease agents. Villemin was part of this experimental environment.

According to his biographers, Villemin had become interested in the contagious properties of tuberculosis after observing the spread of the disease in guards sharing barrack rooms. It has also been suggested that he noticed similarities between the disease and cases of glanders in horses he had encountered whilst growing up on his father's farm. With glanders recognised as contagious and capable of infecting humans, Villemin drew comparisons with bovine tuberculosis and came to the same conclusions.[6] However, veterinarians had commonly regarded glanders as an equine form of tuberculosis until the French dermatologist Pierre-François-Olive Rayer identified that the two diseases were separate. Rayer had also undertaken comparative studies of pulmonary tuberculosis in animals and humans.[7] Villemin had studied with Rayer and it was probably under him that he developed his interest in comparative pathology and tuberculosis. In a series of experimental studies in which purulent liquid from the lung of a patient who had died from tuberculosis was injected into rabbits and guinea pigs, he noted the presence of tubercular lesions after three months. This led Villemin to question the established doctrine of predisposing and hereditary causes. He suggested that tuberculosis had a 'nosological relation' to syphilis, farcy and glanders and hence concluded that it was a specific disease akin to smallpox or typhoid which could be

[5] Lise Wilkinson, *Animals and Disease: An Introduction to the History of Comparative Medicine* (Cambridge, 1992), chs 5 and 7.

[6] S. Lyle Cummins, 'Jean-Antoine Villemin', in *Science, Medicine, and History: Essays on the Evolution of Scientific Thought and Medical Practice written in Honour of Charles Singer*, ed. Edgar Underwood (London, 1953), pp. 331–40.

[7] Wilkinson, *Animals and Disease*, pp. 122–3.

communicated from person to person. The idea that tuberculosis was infective was not new, but Villemin's claims that the disease was caused by the transmission of some disease agent did challenge existing ideas. The pathological results of similar experiments with material derived from a tuberculous cow led him to believe that the disease had a common identity in animals and humans. In putting forward these ideas, he asserted the contagious properties of bovine tuberculosis and cast the disease as zoonotic.[8]

Villemin's methods were not new: inoculation experiments had already started to form the mainstay of pathological studies to determine if a disease was contagious and had been used extensively in French investigations of pleuro-pneumonia.[9] Nor did his findings initially attract much attention when they were first announced to the Académie de Médecine in 1865. However, their public airing at the 1867 Paris Congress created a stir and heightened uncertainties about the pathogenesis and nature of tubercles. Members of the Académie de Médecine were not inclined to accept Villemin's view that tuberculosis was contagious; a similar reaction was encountered in the British medical community, prompting attempts to determine the veracity of Villemin's claims.[10]

So entrenched were ideas of a tubercular diathesis that efforts to reproduce Villemin's results sought to ascribe his success 'to anything rather than infection'.[11] It was commonly felt that animal experiments produced contradictory results and this appeared to be confirmed when Villemin's experiments were repeated. Although studies reaffirmed that injections of 'morbid products' from tubercular patients into animals did result in an 'identical or nearly identical' disease, they suggested that Villemin had produced 'artificial tuberculosis' – that tuberculosis could originate 'traumatically' through injections, a process attributed to a 'common inflammatory' phenomenon.[12] Serious doubts were thus cast on Villemin's conclusions that some form of contagious matter that led to the development of tubercles was being transmitted. As one writer in the *BMJ* concluded:

> The celebrated guinea-pig experiments of Villemin, apparently showing the inoculability of tubercle, and favouring the view of its existence as a zymotic disease, were soon disposed of by the subsequent labours of Drs Andrew Clark, Burdon

[8] 'Causes and Nature of Tuberculosis', *EMJ* xii (1866–7), p. 756.

[9] See 'Pleuro-pneumonia in Cattle', *BMJ* i (1853), p. 549.

[10] For the French reaction, see David S. Barnes, *The Making of a Social Disease: Tuberculosis in Nineteenth Century France* (Berkeley, 1995), pp. 41–7.

[11] George Fleming, 'The Transmissibility of Tuberculosis', *British and Foreign Medico-Chirurgical Review*, Oct. 1874, p. 481.

[12] *Tenth Report of the Medical Officer of the Privy Council for 1867* (London, 1868), p. 20; John Burdon Sanderson, 'Report on the Inoculability and Development of Tubercle', *ibid.* p. 413; Wilson Fox, *On the Artificial Production of Tubercle in the Lower Animals, etc* (London, 1868); *Practitioner* xxix (1882), pp. 178–83.

Sanderson, and Wilson Fox, and we are left, as to the real origin of tuberculous disease, much as we were before.[13]

Despite the reservations expressed by British doctors, and hesitation on the part of the Académie de Médecine, Villemin's conclusions about the relationship between bovine and human tuberculosis prompted further experimenters to test the value of his work as part of an emerging programme of studies into the pathology of contagion and growing interest in animal disease. In the process, bovine tuberculosis emerged as a 'new' disease capable of being transmitted from animals to humans as European pathologists and veterinarians started to ask questions about whether bovine tuberculosis represented a threat to human health, although no one at the time referred to the disease as 'zoonotic'. The link Villemin made between bovine tuberculosis and smallpox, glanders and typhoid, all of which, it was felt, could be spread from animals to humans, provided a focus. Similar findings that asserted both the principle of infection and that bovine tuberculosis could cross the species barrier were reported. The most influential were by Andreas Gerlach, veterinary pathologist and director of the Berlin veterinary school, Edwin Klebs, professor of pathological anatomy at Prague and better known for his discovery of the diphtheria bacillus, and Julius Cohnheim, a German experimental pathologist.[14] Both Klebs and Cohnheim had worked with the celebrated pathologist Rudolf Virchow and were interested in the changes that occurred in animal tissue when it was affected by inflammation, tuberculosis and other disease states. Gerlach was considered a conscientious investigator and made 'the subject of contagious diseases one of special study', having already investigated pleuro-pneumonia and sheep-pox.[15] It was no coincidence that these experimental pathologists were working in Germany, as that country was beginning to supplant France as the centre for new ideas on disease causation.

Gerlach, Klebs and Cohnheim all employed methods familiar to experimental pathologists. All agreed that bovine tuberculosis was contagious, although it was Gerlach's studies on that appeared 'to remove all doubt that [the disease] . . . was identical with Phthisis in mankind'.[16] In defining tuberculosis as a virus, they noted the close 'connection between the virus of the

[13] Richard P. Cotton, 'Notes on Consumption', *BMJ* i (1871), p. 192.

[14] See Daniel H. Cullimore, *Consumption as a Contagious Disease: With its Treatment According to the New Views, to which is Prefixed a Translation of Professor Cohnheim's Pamphlet, 'Die Tuberkulose vom Standpunkte der Infections-Lehre'* (London, 1880); Charles Creighton, *Bovine Tuberculosis in Man: An Account of the Pathology of Suspected Cases* (London, 1881), pp. 22–5.

[15] John Gamgee, 'Cattle Diseases in Relation to Supply of Meat and Milk', *Fifth Report of the Medical Officer of the Privy Council for 1862*, appendix iv (London, 1863), p. 256.

[16] George Fleming, 'Tuberculosis from a Sanitary Point of View', *Veterinary Journal* x (1880), p. 309.

consumptive disease in cattle and that of human tuberculosis'.[17] Through their inoculation experiments, they demonstrated the disease's ability to cross the species barrier. Earlier ideas based around the German histological traditions that bovine and human tuberculosis were not analogous were challenged if not dispelled. In doing so, they questioned Virchow's morphological approach to pathology by asserting the importance of external factors that could provoke disease. This threatened existing conceptions of tuberculosis based on Virchow's cellular pathology that saw the seats of the disease in abnormal changes within cells and tubercles as mere 'knots of tissue' produced by an inflammatory response or abscess formation. Other researchers entrenched in these ideas hence raised questions about the presence of a specific tubercular poison and continued to point to the ambiguous nature of the findings. 'Uncertainty, confusion, and controversial allegations' continued to dominate views of bovine tuberculosis and its contagious properties.[18] This made a consensus problematic.

These European studies were widely reported and discussed in Britain where similar investigations were conducted. These were 'typical', according to Worboys, 'of the speculative germ "discoveries" of the period'.[19] They received official support from John Simon, medical officer to the Privy Council, and built on investigations sponsored by the 1866 Cattle Plague Commission and early work at the Brown Animal Sanatory Institute, using many of the same researchers.[20] Simon sought to promote research into the 'intimate pathology of contagion' in an attempt to trace the 'natural history of contagia' as a branch of what he labelled 'zymotic pathology', a concept that according to Hamlin 'united the new germ theory with the sanitarians' war against filth'.[21] Simon was also keen to extend the regulation of food to help secure the physical welfare of the labouring poor. Interest was further stimulated by the growing anxiety about the health consequences of eating diseased meat, which was reinforced by alarm about the widespread nature of cattle diseases and by the 1865 rinderpest epizootic, which had encouraged renewed investigation into the links between diseased animals, meat and humans. Rinderpest effectively neutralised the unease about pleuro-pneumonia and human health that had characterised earlier concern about meat and animal disease, allowing bovine and human tuberculosis, and the relationship between them, to become the

[17] 'Tuberculosis as a Contagious Disease', *BMJ* i (1880), p. 705.

[18] 'Virchow on the Transmission of Pearl Diseases of Animals by Diseased Meat', *MTG*, 29 May 1880, pp. 582–3.

[19] Worboys, *Spreading Germs*, p. 204.

[20] See Richard Thorne, 'The Effects Produced on the Human Subject by Consumption of Milk from Cows having Foot-and-Mouth Disease', *Twelfth Report of the Medical Officer of the Privy Council for 1869* (London, 1870), pp. 294–9. For the Brown Animal Sanatory Institute, see Wilkinson, *Animals and Disease*, pp. 163–80.

[21] Christopher Hamlin, 'Providence and Putrefaction: Victorian Sanitarians and the Natural Theology of Health and Disease', *Victorian Studies* xxviii (1985), pp. 386–7.

focus of debates about diseased meat. Although pathologists in Britain agreed that the common identity of bovine and pulmonary tuberculosis remained 'anything but proved', a growing number of veterinarians and MOsH started to argue from the 1870s onwards that meat from cattle with tuberculosis represented a threat to human health.[22] Here they were aided by the confusion surrounding the nature of tuberculosis generated by the intense debate over whether it was contagious or hereditary. However, the number of letters to medical journals suggesting that these ideas were not new pointed to a growing current of opinion that favoured an infectious model, and a connection between bovine and human tuberculosis.

Veterinarians and tuberculosis

While general and medical texts concentrated on pulmonary tuberculosis and notions of predisposition and atmospheric stagnation, elite veterinarians in Britain led the way in asserting that bovine tuberculosis was a zoonosis. Despite a general reticence about the value of experimental research, they were quicker to draw practical implications from Villemin's work. With tuberculosis endemic in many herds, veterinarians throughout the 1860s started to become increasingly interested in the disease and evidence that it was contagious in cattle as a way of explaining high levels of infection. Confident about the value of observation, they built on the growing number of European studies and reports that the virus for tuberculosis had been identified – that tuberculosis was infectious and could cross the species barrier. Elite veterinarians borrowed from the experimental and laboratory findings from Europe, particularly those conducted by Gerlach, and added their clinical observations.

In the British veterinary profession, George Fleming spearheaded this interest in bovine tuberculosis, forcing it onto the agenda. Fleming was the son of a shoeing-smith and as a child had worked in the farrier's shop of a veterinary surgeon before entering the service of the Manchester veterinarian, John Lawson, who sent him to Dick Veterinary College in Edinburgh. After qualifying, like many veterinarians keen on advancement, he entered the army veterinary service, serving in the Crimea, China, Syria and Egypt before returning to take up the post of veterinary surgeon to the Royal Engineers. By the mid 1860s, the ambitious Fleming had become vice-president of the Royal College of Veterinary Surgeons (RCVS) and a member of its council. A defender of experimentation and interested in epizootic disease, he was won over by theories that pointed to specific agents as responsible for outbreaks of disease and by continental veterinary practices in general. Fleming endorsed these ideas because he wished to prevent a recurrence of events that had

[22] Reginald Southey, *The Nature and Affinities of Tubercle* (London, 1867), p. vii.

characterised early responses to the cattle plague. He went further, however. Convinced by studies that pointed to the threat posed by diseased meat, and by work on glanders, rabies and anthrax, he argued that certain cattle diseases were transmissible from animals to humans.[23] By asserting the zoonotic properties of animal disease he wanted to bring veterinarians into the 'structure of the British public health organisation' to raise their status and position. As such, Fleming, as an advocate of hygiene, saw an intimate connection between veterinary medicine and public health.[24]

Aware from his own inoculation experiments that tuberculosis was infective, and influenced by European studies that bovine tuberculosis was not limited to cattle, Fleming argued that bovine tuberculosis posed a threat to public health. He first voiced his views in 1874, not in the veterinary press but in the *British and Foreign Medico-Chirurgical Review*, where they would have more impact.[25] In the following year, his *Manual of Veterinary Sanitary Science and Police* emphasised the contagious nature of bovine tuberculosis and its zoonotic properties. Fleming's *Manual* offered a more comprehensive approach to animal health than earlier English veterinary texts, covering the prevention and suppression of epizootic and contagious diseases. It helped strengthen the belief 'that the diseases of man and of the lower animals are much more intimately connected than many authorities seem to suppose'.[26] Over time, Fleming became more forthright, drawing clearer connections between tuberculosis in cattle and the disease in man, arguing that of all the diseases capable of being passed from animals to man, it was 'the most formidable'.[27]

As a leading veterinary army officer Fleming's views proved influential. He was able to use his position as editor of the *Veterinary Journal* to further his ideas on bovine tuberculosis and make them a subject of debate. However, his was not the only voice. Other veterinarians, influenced by anxieties about the role meat played in the transmission of disease, also began to voice unease about the zoonotic properties of bovine tuberculosis and the implications for human health. Thomas Walley, the principal of Dick Veterinary College, had warned of the potential dangers of bovine tuberculosis in 1864 as part of a

[23] 'Contagious Diseases of Animals at the Society of Arts', *Veterinary Journal* ii (1876), p. 283.

[24] George Fleming, *Animal Plagues: Their History, Nature and Prevention* (London, 1871); *idem*, *The Contagious Diseases of Animals; their influence on the . . . Nation, and how they are to be combated, etc* (London, 1876); Anne Hardy, 'Pioneers in the Victorian Provinces: Veterinarians, Public Health and the Urban Animal Economy', *Urban History* xxix (2002), p. 379.

[25] George Fleming, 'The Transmissibility of Tuberculosis', *British and Foreign Medico-Chirurgical Review*, Oct. 1874, pp. 461–86.

[26] *Idem*, *Manual of Veterinary Sanitary Science and Police* 2 vols (London, 1875), vol. ii, pp. 386–90, 393; *British and Foreign Medico-Chirurgical Review* lvi (1875), p. 379.

[27] 'Diseases of Animals Communicable to Man', *Sanitary Record*, 15 Sept. 1880, p. 101.

general treatise on cattle disease.[28] As with Fleming, he was interested in epizootic cattle diseases and their relation to human health. At first his ideas were ill formed and it was not until the early 1870s that he began to state his concerns with more force. Speaking at a veterinary meeting in Glasgow in 1871, Walley was clear that bovine tuberculosis 'was capable of being propagated from animals to man'. Walley had been influenced less by experimental evidence and more by personal tragedy, being certain that he had lost a child to bovine tuberculosis earlier in the year.[29] He was not alone. Encouraged by Fleming, Walley and others, the veterinary press started to express the 'grave suspicion' that tuberculosis was capable of transmission from cattle to humans.[30] As debates about meat and disease increased, leading veterinary manuals in the 1870s pointed to similarities between tuberculosis in humans and the condition in cattle.[31] This did not mean that ideas of heredity were immediately abandoned. Many continued to speak of 'taint[ed]' stock, and suggested that the 'tissues of one particular breed or race are especially favourably disposed to nourish the tubercle bacillus'.[32] However, an increasing number of veterinarians appeared to agree with the *Veterinary Journal* that 'the Tuberculosis of cattle is a transmissible disease, and can be conveyed not only to animals of the same, but also to those of other species in various ways, is now an established fact.'[33]

In reaching this view, many veterinarians were swayed by their encounters with pleuro-pneumonia, foot-and-mouth and rinderpest. Work on glanders in the eighteenth century, and experiences of rinderpest, foot-and-mouth and pleuro-pneumonia in the nineteenth, had seen veterinarians start to investigate the extent to which cattle diseases were contagious. Pleuro-pneumonia and traditional eradication policies to limit epizootics encouraged veterinarians to move away from miasmatic notions of infection and ideas of spontaneous generation that had been an obstacle to introducing quarantine measures. Observational experience and familiarity with parasitic and dietetic diseases reinforced these ideas, as did French studies into how pleuro-pneumonia was spread. The devastating effects of rinderpest, which plunged the cattle industry (and the emerging veterinary profession) into crisis, and the national responses to the disease, served to shock veterinarians into adopting a contagious model that forced a re-evaluation of disease theory, if not an immediate rejection of

[28] Thomas Walley, *The Four Bovine Scourges: Pleuro-pneumonia, foot-and-mouth-disease, cattle plague, tubercle (scrofula)* (Edinburgh, 1879), p. 143.

[29] *Idem, A Practical Guide to Meat Inspection* (Edinburgh, 1896), p. 136.

[30] *Veterinary Journal* viii (1879), p. 196.

[31] See William Williams, *Principles and Practice of Veterinary Medicine* (Edinburgh, 1874), p. 342.

[32] *Departmental Committee appointed to inquire into Pleuro-pneumonia and Tuberculosis in the United Kingdom*, PP (1888) xxxii, p. xxi; *ibid.*, minutes of evidence, p. 88.

[33] 'Can the Milk of Phthisical Cows produce Tuberculosis?', *Veterinary Journal* viii (1879), p. 196.

notions of spontaneous generation. Rinderpest played a further role. Attempts to control the cattle plague provided the veterinary profession with a new public role and raised its status.[34]

By the late 1860s, earlier disagreements about the nature of cattle disease were pushed to one side as most veterinarians came to accept that 'epizootics were contagious and spread by the transmission of some disease-matter'.[35] Support for germ theories grew among veterinarians as they upheld exclusive disease control policies to combat epizootics that reinforced the idea of 'external' or 'foreign' agents that invaded an otherwise healthy system. Outbreaks of epizootics also encouraged a change in veterinary education to include greater attention on the diseases of cattle, sheep and pigs, which helped institutionalise these ideas. Together, these factors stimulated interest in European experimental studies.

With rinderpest and pleuro-pneumonia considered highly contagious, British veterinarians quickly accepted the idea outlined in European studies that bovine tuberculosis was a communicable disease. With tuberculosis difficult to trace in cattle, they were more willing to embrace a model of contagion based on sanitary ideas about hygiene. Influenced by arguments that bovine tuberculosis was contagious, and anxious about its ability to infect other animals based on observational experience that other animals (notably pigs) could also contract the disease, veterinarians began to express apprehension about the implications for human health. Their interest reflected growing concern about questions of food safety and diseased meat and the relationship between animal diseases and illness in humans, highlighted by fears about pleuro-pneumonia and by outbreaks of anthrax and foot-and-mouth. Simple transmission experiments with rabies and glanders from the 1820s onwards had already raised questions about the zoonotic properties of certain animal diseases. In addition, veterinarians could look to other livestock diseases like anthrax, cowpox and foot-and-mouth, which were believed to pass from animals to humans.[36] Work on rabies appeared to confirm that certain animal diseases could cross the species barrier through some form of virus. It helped that bovine tuberculosis in cattle displayed a similar history and pathology to tuberculosis in humans, making the link between the two easier to accept.

In making these claims, veterinarians, as Hardy has argued, were also seeking to expand their range of employment and 'establish the scientific and social

[34] Michael Worboys, 'Germ Theories of Disease and British Veterinary Medicine, 1860–1890', *Medical History* xxxv (1991), pp. 308–27.

[35] Worboys, *Spreading Germs*, p. 44.

[36] See 'Transmissibility of Foot-and-Mouth Disease to Mankind', *Veterinary Journal* ii (1876), p. 213. For glanders, see Lise Wilkinson, 'Glanders: Medicine and Veterinary Medicine in Common Pursuit of a Contagious Disease', *Medical History* xxv (1981), pp. 363–84.

worthiness of their profession'.[37] Veterinary medicine had a low status and elite veterinarians were struggling to define their professional domain and to escape the tarnish of the farrier's trade.[38] The RVC was not established until 1791 and until the 1840s was beset with problems, dominated as it was by Edward Coleman, who had little knowledge of veterinary medicine. The profession was not formally unified until the passage of the 1881 Veterinary Surgeons Act, whilst the private veterinary schools were reluctant to concede the RCVS any real authority over qualifications. Although the RVC offered classes on livestock epizootics, training was dominated by equine studies and the health of farm animals was neglected in an environment in which most veterinary practices were constructed around the horse.[39] Despite attempts to improve academic standards and to widen the breadth of the curriculum to include classes on cattle pathology, farmers and cowkeepers had little confidence in veterinary surgeons, noting that they 'have nothing to do with cattle diseases'.[40] The inability of veterinarians to protect Britain from cattle diseases only appeared to confirm this assessment. Veterinary surgeons wanted to assert their area of competency and define themselves as 'scientific' practitioners to boost their position, especially as early nineteenth century studies of animal diseases had been dominated by doctors. Elite veterinarians therefore borrowed a medical and administrative language that had parallels with the public health movement as they asserted their credentials to speak on animal health issues when they related to public health. These ideas were voiced with increasing frequency throughout the 1870s and were accompanied by efforts by veterinarians to move into local government and public health at a time when their role was expanding following the 1878 Contagious Diseases (Animals) Act. By identifying bovine tuberculosis as a danger to humans, and by insisting on their ability to detect the disease in living cattle and dead meat, elite veterinarians were seeking to move into public health work.

[37] Hardy, 'Pioneers', p. 380.
[38] John R. Fisher, 'Not Quite a Profession: The Aspirations of Veterinary Surgeons in England in the Mid-Nineteenth Century', *Historical Research* lxvi (1993), pp. 284–302.
[39] See Iain Pattison, *The British Veterinary Profession, 1791–1948* (London, 1984); Roy Porter, 'Man, Animals and Medicine at the Time of the Founding of the Royal Veterinary College', in *The Advancement of Veterinary Science: The Bicentenary Symposium Series*, vol. iii, *History of the Health Professions: Parallels between Veterinary and Medical History*, ed. A. R. Michell (Oxford, 1993), pp. 19–30.
[40] *Royal Commission on the Cattle Plague*, PP (1865) xxii, q. 3823–4; John R. Fisher, 'The European Enlightenment, Political Economy and the Origins of the Veterinary Profession in Britain', *Argos* xii (1995), p. 47.

Doctors, contagion and bovine tuberculosis

Although *The Lancet* accused veterinarians of apathy, it was a charge Walley felt 'might be lodged with much greater justice at the door of the medical profession' when it came to bovine tuberculosis.[41] It was not until the late 1870s that a growing number of doctors began to be swayed by veterinarians' arguments that bovine tuberculosis was communicable to humans. They were later to claim that it was the medical profession that had identified the danger.

Part of the explanation why doctors were slow to see the public health risks of bovine tuberculosis rests in the nature of veterinary medicine in the mid nineteenth century. As already noted, veterinary medicine suffered from a low status and a meagre and restricted training. Doctors regarded veterinary medicine as a 'dirty, black-looking trade' dominated by those who 'love animals for their own sake, and wish to be always amongst them'.[42] The role surgeons played in examining students for the RVC cast the medical profession as possessing a superior body of knowledge that helped reinforce the impression that veterinary surgeons were subordinate to their medical colleagues, a situation that continued at the Brown Animal Sanatory Institution, the major centre for research into comparative medicine in Britain. It did not help that veterinary sanitary science was in its 'infancy'.[43] In addition, fearful of competition, doctors strenuously resisted any encroachment of veterinarians into human medicine, especially as veterinary surgeons' concentration on exclusive disease control policies contrasted with the limited means available to MOsH to combat contagious disease. Doctors were therefore initially sceptical of the findings put forward by veterinarians and, as Walley told the National Veterinary Association in 1883, were slow 'to take anything in hand which is suggested by' veterinarians.[44]

However, it was not just a question of status. Confusion surrounded the nature of tuberculosis. Medical views about the possible contagiousness of the disease remained vague and contradictory. Notions of constitutional predisposition and heredity dominated the medical understanding of tuberculosis (and other diseases) because 'it made sense of the otherwise inexplicable distribution of the disease'.[45] As a hereditary disease, tuberculosis was a constitutional and idiopathic condition and hence non-preventable. As such, it was not a public health issue. This view only started to be seriously questioned in the 1860s. At the same time, despite the importance of animal models to experimental physiology and pathology, and the early interest taken by the

[41] *BMJ* i (1888), p. 419.
[42] *EVJ* iii (1861), p. 698.
[43] 'Veterinary Sanitary Science', *British and Foreign Medico-Chirurgical Review* lvi (1875), p. 378.
[44] *BMJ* i (1888), p. 419.
[45] F. B. Smith, *The Retreat of Tuberculosis, 1850–1950* (London, 1988), p. 37.

MAMOH in diseased meat, it was only following the cattle plague that a growing number of doctors began to become seriously troubled by animal diseases and food safety. That man could be viewed as a 'creature *sui generis*' initially contributed to a sense that human and animal disease were separate.[46] Comparative pathology had a low status and, along with zoology, was not part of mainstream medical education. The institutional environment for experimental and comparative medicine in Britain, unlike continental Europe, was limited and the laboratory accommodation available in medical schools was often confined to inadequate facilities. Experimental studies were considered secondary to clinical observation, and laboratory investigations were viewed with hostility. Experimental work on tuberculosis therefore belonged to an area of investigation that many doctors viewed with scepticism.

It was only from the 1870s when a new breed of experimentalists began to turn their attention to a comparative approach to medical problems that concern about bovine tuberculosis grew. As interest in the mechanisms of infection was refined, calls were made for a study of animal diseases as a way of throwing light on contagion and the successful control of epidemic disease. This approach was encouraged by Simon at the Privy Council. Doctors, swayed by the evidence supplied by studies for the Privy Council and its successor the Local Government Board (LGB), were beginning to suggest that zymotic diseases were caused by 'an invasion of the animal body by a distinct extremely minute, living, and self-multiplying thing'.[47] Although confusion remained as to what germs were, doctors began to talk about the 'germ of the disease' with more confidence, so that by the mid 1870s experimental studies based around 'parasitic-germ theory' were starting to challenge clinical observation in the development of etiological theories of disease. Work on animal and cattle diseases was part of this study of 'morbific contagia' and the search for the agents responsible.[48] It was felt that laboratory work with animals provided a reference point for understanding contagion. More support was therefore voiced for comparative pathology as the idea gained ground that 'the pathology of domestic animals could not be wholly or properly separated from that of man' and that the 'zymotic diseases of domestic animals' were 'governed by the same general laws as the corresponding diseases of men'.[49]

Although studies of smallpox had already suggested that certain animal diseases could be transmitted to humans, and work on glanders provided a reference point for a large body of investigations into zoonotic disease, medical interest in animal diseases was revived by outbreaks of foot-and-mouth disease and the fears that came to surround rabies. Articles on veterinary issues and

[46] MTG, 29 May 1880, p. 582.

[47] *Annual Report of the Medical Officer of the Local Government Board for 1877* (London, 1878), p. 169.

[48] See Worboys, *Spreading Germs*.

[49] *BMJ* ii (1880), p. 473; William Farr, *Vital Statistics* (Metuchen, NJ, 1975), pp. 328–30.

epizootics started to appear in the medical press, and doctors began to write about animal diseases with greater frequency. Edward Klein, 'the father of bacteriology' in Britain, was at the forefront of this work. Initially at the Brown Animal Sanatory Institute and then at St Bartholomew's Hospital, he undertook investigations for LGB that explored the relationship between animal and human disease. Working from a histo-pathological background, Klein employed comparative pathology and animal experiments to shed light on the process of infection.[50] Although the restrictions on vivisection imposed by the 1872 Cruelty to Animals Act did slow development, this renewed enthusiasm for comparative pathology stimulated investigation into bovine tuberculosis and epizootics as part of the growing concern with the identity of contagion.[51]

Against this background, questions were raised about the nature of tuberculosis, its unitary character and cause. Mortality from tuberculosis had risen in the mid 1860s, which helped concentrate attention on the disease. Although claims by Villemin and other veterinarians that tuberculosis was contagious had been dismissed by doctors as mistaken, the idea that tuberculosis was a communicable disease started to attract growing attention in medical circles in the 1870s. Bovine tuberculosis fitted well with these debates.

The possibility that bovine tuberculosis could be spread from animals to humans was first mooted in the medical press in 1876. The 1875 Public Health Act encouraged a dramatic rise in the number of convictions for the sale of diseased meat, exciting interest in the relationship between animal and human disease. Outbreaks of pleuro-pneumonia further heightened concern. MOsH looked for guidance, aware that 'the question of the fitness of meat for food' was in 'an unsettled state'.[52] As part of this renewed debate on what meat was fit for human consumption, some MOsH began to hint at the possibility that bovine tuberculosis was capable of being spread from animals to humans, although they at first made little reference to veterinary studies. For example, in a case against Thomas Bulmer, a butcher in Warrington found to be stewing diseased meat in 1880, the MOH called to give evidence claimed that 'persons eating the stew and meat were liable to be affected with a similar disease (tuberculosis) to that of which the cow had died'.[53] Others were more cautious. Scepticism about bovine tuberculosis as a 'contagious malady' and its relationship to the disease in humans continued to be voiced. In an address to the

[50] Edward Klein, 'Report on the Lymphatic System of Skin and Mucous Membranes', *Annual Report of the Medical Officer of the Local Government Board for 1879* (London, 1880), pp. 102–34; 'Tabular Statement showing Results of Experiments by Dr Klein in the Inoculation of Bovine Animals with Smallpox', *ibid.*, pp. 135–42.

[51] See Richard D. French, *Antivivisection and Medical Science in Victorian Society* (Princeton, NJ, 1975).

[52] Junius Hardwicke, 'Meat as Food for Man', *Sanitary Record*, 15 Feb. 1880, p. 284.

[53] Bulmer was found guilty and fined £32 15s: 'Shocking Charge of Selling Bad Meat', *Meat and Provisions Trades' Review*, 31 Jan. 1880, p. 710.

Northern and Yorkshire Associations of Medical Officers of Health in 1879, Francis Vacher, MOH for Birkenhead, acknowledged attempts to show that human and bovine tuberculosis were the same, but explained that the evidence was not convincing. Like many doctors, he gave little credit to the ideas put forward by veterinarians.[54] The British Medical Association (BMA) at its Cambridge meeting in 1880 further discussed the issue. The meeting revealed conflicting opinions though most who spoke questioned whether bovine tuberculosis was communicable from animals to humans. It called for further investigation having given short shrift to Fleming.[55]

Just as Vacher was questioning the possibility that bovine tuberculosis was a communicable disease, interest in the disease in medical circles was encouraged by the pathological findings of Charles Creighton, demonstrator of anatomy at the University of Cambridge. He was a controversial figure given his stance in the anti-vaccination movement and had already made a name for himself by pointing to errors in Klein's work on sheep-pox.[56] First in *The Lancet* in 1880, and then in *Bovine Tuberculosis in Man* the following year, Creighton stated that some cases of tuberculosis in man had been 'derived from the bovine'. Although influenced by Gerlach's experiments, he wanted to assert the role of doctors in defining the disease. Central to his work was the question whether the disease could also be 'communicated to man'.[57]

For Creighton the answers could not be found 'by tracing the individual cases to particular sources of poisoning or of infection', nor 'by the test of experimentally inducing the disease by the inoculation of the substance in animals'. Instead, he relied on post-mortem evidence from twelve patients who had died from tuberculosis at Addenbrooke's Hospital in 1880. Creighton placed his faith in pathological anatomy. He was defending a British approach to the classification of disease based on solid anatomical work and clinical experience. In doing so, he paradoxically rejected inoculation experiments despite his use of studies that drew on them. What was important for Creighton was a style of morbid anatomy and observational tradition that remained strong in English medical schools. The 'morphological resemblances' and '*structural mimicry*' of the tubercles he observed between human and bovine tuberculosis during his microscopic studies were enough to convince him that tuberculosis in cattle and humans were related.[58]

[54] *Sanitary Record*, 20 June 1879, p. 289.
[55] 'Diseases of Animals Communicable to Man', *ibid.*, 15 Sept. 1880, pp. 101–2.
[56] *Report of the Medical Officer of the Privy Council and Local Government Board* (London, 1876), p. 23; Charles Creighton, 'Note on Certain Unusual Coagulation Appearances found in Mucus and other Albuminoidal Fluids', *Proceedings of the Royal Society of London* xxv (1877), pp. 140–4.
[57] Charles Creighton, 'An Infective Variety of Tuberculosis in Man, Identical with Bovine Tuberculosis', *Lancet* i (1880), pp. 943–94.
[58] Creighton, *Bovine Tuberculosis*, pp. 4, 6, 5.

Creighton regarded the twelve cases he had examined as by no means infrequent. He went on to surmise that there were probably many more cases that had escaped detection because they 'have not been called bovine tuberculosis'. His suspicions that the disease could change its character, led him to suggest that certain cases of typhoid might also be attributed to bovine tuberculosis. He pointed to the 'readiness with which bovine tuberculosis may be transmitted'. To support this view he drew parallels with glanders and the belief that the disease could be transmitted from horses to humans.[59] Only later was he to express doubts.[60]

Creighton's work attracted attention, aided by his talent for self-publicity. The *Glasgow Medical Journal* felt that he had conclusively demonstrated that cases of bovine tuberculosis did 'occur in man'.[61] Even Vacher shifted his stance, noting in 1881 that 'the weight of recent pathological evidence tells strongly in favour of the identity of the human and bovine disease'.[62] Although not all were won over, with one review in the *Sanitary Record* commenting that Creighton had failed to offer sufficient proof, his work was used by other doctors when seeking to understand the risk of infection from meat from tuberculous cows.[63]

At the forty-eighth annual meeting of the BMA in Cambridge, papers on bovine tuberculosis by Vacher and Fleming prompted the resolution that: 'the communicability of disease to man by animals used by him as food urgently demands careful inquiry, both in regard to the actual state of our knowledge thereon, and to the legislation which is desirable in connection'. Attention was increasingly turning to specific points of passage in the transmission of disease and the role farm and domestic animals played as sources or carriers. Claims that bovine and human tuberculosis were related were used to support proposals for further study into 'the pathology of domestic animals' based around the argument that it should 'not be wholly or properly separated from that of man'.[64] Although the 1881 edition of *Histologie Pathologique* noted that 'the question of tuberculosis' continued to bristle 'with contradictory opinions', the weight of pathological evidence pointed to a common identity between human and bovine tuberculosis and its contagious properties.[65] Within a year, the announcement by Koch to Berlin Physiological Society that he had identified the bacillus responsible recast debate.

[59] *Ibid.*, pp. 1–3, 10, 103, 101.
[60] Charles Creighton, *Contributions to the Physiological Theory of Tuberculosis* (London, 1908), p. 157.
[61] GMJ xvi (1881), p. 286.
[62] Frances Vacher, *What Diseases are Communicable to Man from Diseased Animals Used as Food?* (London, 1881), p. 19.
[63] *Sanitary Record*, 15 Jul. 1881, p. 30; *Transactions of the Liverpool Medical Institution* (1881–2), p. 103.
[64] *BMJ* ii (1880), p. 473.
[65] *BMJ* i (1881), p. 818.

'The bacillus, microbe and tubercle age'

As with many other pathologists at the time, Koch was looking for the 'necessary' cause of disease using a system of cultures and inoculation experiments that drew on comparative pathology.[66] In the process he made a number of technical innovations in microscopy, experimental pathology and culturing, making the latter less cumbersome and less likely to produce contamination. Koch's work is associated with the 'Golden Age of Bacteriology' and, with Pasteur, he is seen as one of the principal fathers of this new science. However, as Worboys has shown, the situation was more complex.[67]

Already known for his identification of the anthrax bacillus in 1876 whilst working as a country practitioner, Koch moved to Berlin in 1880 as a salaried scientist in the Imperial Department of Health. Here he continued to refine his bacteriological methods and in 1882 started research on tuberculosis. Within eight months, he announced his discovery of the tubercle bacillus, and in doing so asserted that it was contagious. Koch's achievements were considerable, since the bacillus was difficult to grow. Having identified the micro-organism responsible, Koch also suggested that 'bovine tuberculosis is identical with human tuberculosis'. He went on to argue that bovine tuberculosis was 'undoubtedly [a] . . . source of infection' in humans. Koch added that if it was probably less virulent than the human form of the disease 'it must by no means be underrated'.[68] Koch based his ideas on a unified model of tuberculosis as an infectious disease that encompassed both its bovine and human forms and owed much to the ideas of Laënnec and Villemin. Koch's claims were initially modest and the attention he attracted was, as Rosenkrantz has argued, 'more from style than substance'.[69] However, when he published his second and more expansive article on the 'Aetiology of Tuberculosis', Koch was clear that tuberculosis in humans and in animals *was* caused by the same specific bacillus.[70]

Worboys has suggested that Koch's discovery was not a major discontinuity in 'theories' of consumption; that the major novelty was the issue of contagion.[71] Koch himself admitted that his view on the infectious properties of tuberculosis 'scarcely offers anything new'.[72] The notion that tuberculosis

[66] For a balanced account of Koch's work, see Thomas D. Brock, *Robert Koch, a Life in Medicine and Bacteriology* (Madison, WI, 1988).

[67] See Worboys, *Spreading Germs*.

[68] Cited in Barbara G. Rosenkrantz, 'The Trouble with Bovine Tuberculosis', *BHM* lix (1985), p. 156.

[69] *Idem*, 'Koch's Bacillus: Was there a Technological Fix?', in *The Prism of Science* 2 vols, ed. Edna Ullmann-Margalit (Boston, 1986), vol. ii, p. 151.

[70] 'Nach einem in der Physiologischen Gesellschaft zu Berlin am 24. März cr. gehaltenen Vortrage', *Berliner klinische Wochenschrift*, Bd. xix (1882), pp. 221–30.

[71] Worboys, *Spreading Germs*, p. 193.

[72] 'Ought Koch's Tubercle-Pathology to be Accepted', MTG, 29 Apr. 1882, p. 441.

might be caused by a virus was already well rehearsed by the 1870s, although the heredity framework proved hard to dislodge. Nor did Koch's statements on bovine tuberculosis represent a break with existing thinking. Although Koch's identification of the tubercle bacillus did assert the value of bacteriology in defining bovine tuberculosis and the idea of contagion in medical discourse, inoculation experiments and pathological observations in Europe and Britain had already pointed to the suggestive likeness between tuberculosis in humans and cattle. As *The Lancet* explained, the idea that bovine tuberculosis was 'the analogue in the bovine species of tuberculosis in the human' had 'long been admitted' before 1881.[73] Studies had acknowledged the role of specific organisms in the transmission of disease, and that bovine tuberculosis was contagious and could be spread to humans. Cohnheim predicted the discovery of an 'infective agent' for tuberculosis in 1879 and two years later in a paper to the Paris Académie des Sciences, Henri Bouley talked about the 'germ origins of tubercle' in cattle. These findings had been widely reported.[74] Even Koch admitted that his infective theory was not new, 'it has only been confirmed by these researches'.[75] For the *Journal of Comparative Pathology and Therapeutics*, Koch merely 'completed in the most happy manner the demonstration of the infectious nature of tuberculosis'.[76]

However, because of his status and earlier work on anthrax, Koch did bring an 'unanswerable confirmation to the theory of the unit of tuberculosis, and of the communicability of the disease from animals to man'.[77] Although initial reactions to the tubercle bacillus were mixed, Koch's experiments encouraged doctors to view the disease in increasingly ontological terms. He gave authority to studies made by veterinarians and made the ideas embedded in them more palatable to doctors. Several factors aided this process. First, Koch published his findings during a period of experimental interest in bovine tuberculosis. Second, his ideas 'arrived at a time when the identification of possible causative agents of contagious, infectious and other diseases was becoming routine'.[78] Third, they were bolstered by a growing acceptance that many diseases common to man and the lower animals were communicable from one to the other following work on rabies, glanders and anthrax. Finally, they were made at a time when debates about animal disease and meat were increasing in intensity.

As discussions in medicine concentrated on specific points of passage, and as growing interest focused on animal diseases, studies were undertaken into bovine tuberculosis that were rooted in experimental pathology and the new

[73] *Lancet* i (1882), pp. 655–6.
[74] Creighton, *Contributions*, p. 7; *MPC*, 5 Oct. 1881, pp. 301–2; 'Human and Bovine Tuberculosis', *Veterinary Journal* i (1882), p. 342.
[75] 'Koch on Tuberculosis', *ibid.* ii (1884), p. 38.
[76] 'Congress for the Study of Tuberculosis in Man and Animals', *JCPT* i (1888), p. 263.
[77] Sheridan Delépine, 'Tuberculosis and the Milk Supply', *Journal of Meat and Milk Hygiene* i (1911), p. 545.
[78] Worboys, *Spreading Germs*, p. 211.

science of bacteriology to confirm the relationship between the tubercle bacilli and all forms of tuberculosis.[79] Koch's work with animals made inoculation experiments an ordinary laboratory routine and helped define the new science of bacteriology. Practices were standardised and Koch's work was repeated to determine its veracity by a wide range of practitioners. Many of these studies never appeared 'in literature because, being simply confirmatory of Koch's results, their communication was considered unnecessary'.[80] This desire to confirm Koch's findings arose from a wish to develop a natural history of bacteria, as doctors, caught up in a wave of enthusiasm for the potential of bacteriology, sought to catalogue the agents responsible for the major infectious diseases. As Worboys has explained, the tubercle bacillus was used 'as a vehicle for the wider promotion of bacterial theories of disease'.[81] Standardised procedures and the do-it-yourself nature of laboratory work, along with better training, made this easier. Interest therefore was not limited to a small group of pathologists working for the LGB or MOsH. If the tubercle as an infective disease had been contested, tubercles in their bacilliary form were accepted surprisingly quickly. The bacillus, therefore, swiftly found its way into texts on pathology, medicine and veterinary medicine. By 1884, as one veterinarian explained, 'you cannot take up a professional *journal*, either medical, veterinary, or agricultural, but you find "*Tubercle*" staring you in the face'.[82] If not all were persuaded, the idea that tuberculosis was contagious gained ground, especially as this conception of the disease appeared to promise new treatments. In public health terms, it also assured the possibility of targeting disease agents and nuisances with greater precision, and in doing so of improving the efficiency of public health medicine.

Inspired by Koch's claims, doctors started to defend their role in the identification of bovine tuberculosis as a zoonotic disease as the subject 'assumed an importance which it did not possess a very few years ago in the eyes of scientific men'.[83] Veterinarians' commitment to clinical observation was used against them and doctors asserted medicine's role in defining bovine tuberculosis as a public health issue. Concerns about the health of the poor and the health of the child, in large part due to the persistence of high infant mortality rates, were voiced with greater frequency, as doctors began to employ pathological and bacteriological findings to explain the threat of bovine tuberculosis to health. Support was given to the link between the ingestion of tuberculous matter from cows and cases of tuberculosis in humans, as debates

[79] See *Annual Report of the Medical Officer of the Local Government Board for 1886* (London, 1887), pp. 415–20; *Annual Report of the Agricultural Department of the Privy Council on Contagious Diseases, etc., for 1888* (London, 1889), appendix.

[80] Alexander Wynter Blyth, A *Manual of Public Health* (London, 1890), p. 457.

[81] Worboys, *Spreading Germs*, p. 206.

[82] A. MacGillivray, 'Tuberculous Milk', *Veterinary Journal* ii (1884), p. 309.

[83] *Departmental Committee . . . into Pleuro-pneumonia and Tuberculosis*, minutes of evidence, p. 284.

about the contagious properties of tuberculosis made the disease both a medical and a public health issue.

Throughout the 1880s a consensus about the pathogenic nature of bovine tuberculosis grew, fuelled by fears about rising levels of tuberculosis in the population and by debates on the dangers of eating diseased meat. Although the evidence was often circumstantial, it served to reinforce the idea that certain diseases in humans and cattle were 'alike' and dangerous, reflecting mounting anxiety about bovine tuberculosis and food safety.[84] A growing number of veterinarians, bacteriologists, physicians and sanitarians were prepared to endorse experimental work into bovine tuberculosis. Slight variations between the two bacilli were overlooked by researchers won over by the 'bacilliary doctrine of tuberculosis' that came to dominate 'the scientific world' by the 1890s.[85] Opinion shifted to favour bovine and human tuberculosis to be one and the same disease. As one doctor noted

> We find that in recent days the majority, if not the whole of those who have examined the matter most closely, agree to the infectiousness of the disease. Looking at it from this point of view, we are bound to pronounce it one of the most serious maladies which affect animals, not simply because of its insidiousness and vitality, but from the fact of its being transmissible from the bovine to other species, and from animals to man.[86]

This contagious model did not mean that hereditary notions were entirely displaced. Those involved in the institutional care of consumptive patients remained sceptical, in part because a contagious model raised difficult questions about grouping tuberculous patients together in a hospital setting. Ideas that infection depended on 'different soil' were used to explain why cases 'derived from the bovine' were higher amongst the poor.[87] However, most doctors and veterinarians seemed increasingly willing to accept that 'there can be no question that Tuberculosis is transmissible from the bovine to the human species'.[88]

Growing fears about bovine tuberculosis led the Privy Council to take an interest in the subject. In 1888, it appointed a departmental committee to investigate measures to eliminate pleuro-pneumonia following concerns about the rising levels of the disease, the cost to the farming industry, and suggestions

[84] *Annual Report of the Medical Officer of the Local Government Board for 1886*, pp. xiii–vii.

[85] *Lancet* ii (1889), p. 271; 'Important Trial Regarding Tuberculous Carcases at Glasgow', *JCPT* ii (1898), pp. 188–9.

[86] Cited in *Tuberculous Meat: Proceedings at Trial under Petitions at the Instance of the Glasgow Local Authority against Hugh Couper and Charles Moore* (Glasgow, 1889), p. 268.

[87] *The Times*, 27 Oct. 1889, p. 9.

[88] August Lydtin, George Fleming and M. van Hertsen, *The Influence of Heredity and Contagion on the Propagation of Tuberculosis and the Prevention of Injurious Effects from Consumption of the Flesh and Milk of Tuberculous Animals* (London, 1883), p. 168.

that inoculation was an alternative to traditional slaughter policies.[89] The departmental committee included bovine tuberculosis in its deliberations after becoming alarmed by the frequency of the disease in cattle slaughtered under 'pleuro-orders'. It was also under pressure from butchers, meat traders and sanitary authorities to include the disease within its remit.[90] With so many investigations into bovine tuberculosis now asserting its contagious properties, these groups wanted guidance on the threat posed by the disease to humans. Although the final report of the committee mainly dealt with pleuro-pneumonia and concentrated on measures to protect cattle, it did endorse the view that bovine tuberculosis could 'be transmitted to man from the lower animals', explaining that it was responsible for 10–14 per cent of 'deaths among human beings'. It therefore outlined proposals for the prevention of infection and the eradication of the disease. These built on existing public health legislation and the need to include tuberculosis under the Contagious Diseases (Animals) Acts, a suggestion that reflected the standard response of elite veterinarians to the threat of livestock diseases.[91]

International congresses also turned their attention to bovine tuberculosis. Delegates to the 1888 Paris congress on tuberculosis called for every possible means to be taken to prevent infection from the disease. The congress had an important impact on British doctors and veterinarians. It helped set the international agenda and demonstrated how far the understanding of bovine tuberculosis had changed during the preceding decade.[92] Direct links were made between the control of the disease in cattle and attempts to combat the spread of tuberculosis in humans. The same concerns were reflected in the popular press. *The Times* noted in 1889 that 'there is no doubt at all that tubercle can be communicated from one animal to another' and between animals and humans.[93] These ideas could gain acceptance in part because bovine tuberculosis could be fitted into growing efforts to prevent the spread of tuberculosis. At the same time, bacteriological interpretations could be merged with earlier ideas about heredity, seed and soil. As a result, bovine tuberculosis came to be cast as one of the greatest pests of humanity and the animal kingdom.

[89] *Departmental Committee . . . into Pleuro-pneumonia and Tuberculosis*, p. xi.
[90] GCA: 'Tuberculosis in Cattle', Mar. 1888, MP17.309.
[91] *Departmental Committee . . . into Pleuro-pneumonia and Tuberculosis*, pp. xxiii, xxiv.
[92] 'Communicable Diseases Common to Man and Animals and their Relationship', *Sanitary Record*, 15 Sept. 1888, p. 122.
[93] *The Times*, 27 Oct. 1889, p. 9.

Conclusions

Despite growing support in veterinary circles in the 1860s and 1870s for the idea that tuberculosis was inter-communicable, it was Koch's claims in 1882 that bovine and human tuberculosis were caused by the same micro-organism that offered doctors conclusive proof. Koch's work not only provided a basis for creating the discipline of bacteriology, his germ theory changed the language in which bovine tuberculosis was described, even if it offered little that was new. The identification of the tubercle bacillus produced a certainty that had previously been lacking that the products from diseased livestock were dangerous. As a result, it was quickly embraced by MOsH because it offered a persuasive rationale for the control of the meat trade. The disease was able to generate concern not only because it was clearly identified as infectious, thereby offering conclusive evidence that diseased meat was dangerous to health, but also because it was felt to be 'much more disastrous in its results than all the other infectious diseases put together'.[94] Manuals directed at public health officials and at veterinarians started to devote whole sections to the disease and the implications of consuming products from tubercular livestock. Interest was focused by rising anxiety about the nature of meat supply and where this meat came from, particularly in the light of concerns about foreign competition. In the process, diseased meat became synonymous with tuberculous meat.

[94] Henry Behrend, *Cattle Tuberculosis and Tuberculous Meat* (London, 1893), p. 86.

4

'No inconsiderable danger to the community'

Meat and bovine tuberculosis

According to *The Times* in 1889, tuberculous meat had come to represent 'no inconsiderable danger to the community'.[1] The *Medical Press and Circular* added that selling such meat was little short of 'constructive murder' for 'there was no telling how many lives might be sacrificed'.[2] Throughout the 1880s, the medical and veterinary press reported widespread alarm at the amount of tuberculous meat that the public was being duped into buying. MOsH and veterinary surgeons repeatedly warned of the dangers of eating meat from cattle with tuberculosis whereas previously they had spoken in general terms about diseased meat and the threat to public health.

According to Perren this concern about bovine tuberculosis reflected more 'general anxiety' about urban public health.[3] However, anxiety went deeper than general unease: it reflected mounting alarm about the relationship between meat and disease, between epizootics and meat, that had been growing throughout the 1850s and 1860s. As the quantity of diseased meat sold was seen to rise, bovine tuberculosis moved to the heart of these debates and provided a focus for existing fears about diseased meat as sanitarians sought to clean up the meat trade. Whereas studies of the dangers of meat from cattle suffering from pleuro-pneumonia, rinderpest, anthrax and foot-and-mouth in the 1850s and 1860s had minimised the public health risk, by the 1880s tuberculous meat had emerged as the archetypal diseased meat. It stood in the 'front rank of importance' in discussions of animal disease because, unlike other zoonotic diseases, the train of transmission was confirmed by experimental studies.[4] However, whereas putrid meat was relatively straightforward to spot, tuberculous meat created far more problems. What was meant by tuberculous meat and the extent of the danger proved hard questions to answer.

[1] *The Times*, 27 Oct. 1889, p. 9.

[2] MPC, 28 June 1882, p. 556.

[3] Richard Perren, *The Meat Trade in Britain, 1840–1914* (London, 1978), p. 53.

[4] *Practitioner*, Nov. 1889, p. 395.

'One man's meat is another man's poison'

Suspicions about the dangers of meat from tubercular livestock were expressed shortly after Jean-Antoine Villemin's announcement in 1865 that the bovine and human forms of the disease were related and that tuberculosis was contagious. A number of pathologists began to examine how the disease might be spread from cattle to humans as part of their investigations into the 'pathology of contagion'. Edwin Klebs, professor of pathological anatomy at Prague, was among the first to hypothesise that it entered the body 'in the great majority of cases by way of the digestive tract'.[5] It was a view that fitted with notions that tuberculosis was a poison or virus that was spread through the blood. The French veterinarian Jean-Baptiste Chauveau came to similar conclusions in a paper to the Académie de Médecine in 1868.[6] He had developed an interest in the comparative pathology of infectious disease following his work on smallpox and turned to animals as a veterinarian working in the busy veterinary school at Lyon. Much of this work examined the communicability of various infections through the digestive system. Chauveau suggested that bovine tuberculosis in humans could be 'propagated with ease by the consumption of meat' from infected cattle. The paper created a storm but, since Chauveau's study was based on a small sample, it was considered inconclusive.[7] Other researchers, including the leading German pathologist Rudolf Virchow, sought to dismiss such findings, adamant that 'no man has yet acquired [tuberculosis] through partaking of tuberculous flesh'. However, a growing body of experimental evidence did begin to point to a different interpretation that contested the notion that tuberculosis was a form of inflammation.[8]

Whereas Klebs, Chauveau and other European pathologists proposed that meat could transmit bovine tuberculosis, it was Andreas Gerlach's work at the Berlin veterinary school that asserted the 'transmissibility' of the disease 'through the digestive organs' that attracted most attention. Gerlach had started his feeding experiments in an attempt to clear up the question of whether humans incurred any risk from eating meat from tuberculous cattle. This followed on naturally from his studies that linked the bovine and human forms of the disease. After conducting more than 110 experiments, he concluded that the 'flesh of a tuberculous cow is . . . infectious'.[9] Evidence that other diseases, notably anthrax, rinderpest and scarlet fever, could be spread

[5] *Archiv für pathologische Anatomie und Physiologie und für klinische Medizin* xliv (1868), p. 266.

[6] August Lydtin, George Fleming and M. van Hertsen, *The Influence of Heredity and Contagion on the Propagation of Tuberculosis and the Prevention of Injurious Effects from Consumption of the Flesh and Milk of Tuberculous Animals* (London, 1883), pp. 113–14.

[7] MTG, 28 Nov. 1868, p. 621.

[8] Cited in Charles Creighton, *Bovine Tuberculosis in Man: An Account of the Pathology of Suspected Cases* (London, 1881), p. 10.

[9] 'Transmissibility of Tuberculosis', *Veterinary Journal* v (1877), pp. 203–5.

by meat from diseased cows, and that in pleuro-pneumonia and sheep-pox the organs affected had 'contagious properties', only seemed by analogy to uphold suspicions about bovine tuberculosis.[10]

Although experimental studies pointed to the role of meat in the spread of bovine tuberculosis to humans, initially there was confusion about the extent of the danger. Studies by other European pathologists put forward a less favourable interpretation. Whilst they admitted that eating meat from 'consumptive subjects' could lead to tuberculosis, they argued that this only occurred when the cow was in an acute stage of the disease. Others were more forthright. They sought to downplay the danger presented by meat, fearful of the consequences for the meat trade, and for the poor, given the large amount of tuberculous meat that was sold in most urban markets.[11] Comparisons were drawn between meat from tuberculous livestock and meat from cattle suffering from rinderpest and pleuro-pneumonia, which had not proved dangerous. It was argued that bovine tuberculosis was probably the same. Those doctors who put forward this view asserted that it was unfair to deprive the poor of those meagre portions of meat that they could afford, and reasoned that common-sense suggested that tuberculous meat was rarely pathogenic. It was argued that because a large proportion of cattle were diseased it was difficult to avoid eating tuberculous meat, and that most people had consumed it without adverse consequences.[12] Butchers were keen to support this assessment. By playing down the risks of eating meat from cattle with tuberculosis, the need for action was reduced and the meat industry protected.[13] However, such evidence did not meet the needs of those who wanted greater regulation of the meat trade.

Despite these doubts, the belief that bovine tuberculosis was 'probably more frequently conveyed through meat than was supposed' gradually gained acceptance, as the amount of meat from tuberculous livestock that was sold increasingly caused alarm.[14] As studies pointed to the fact that bovine tuberculosis was a zoonotic disease, meat became an obvious suspect in the search for how the disease was passed to humans. Edmund Parkes, professor of military hygiene at the Royal Victoria Hospital at Netley, had already established the principle that 'all diseases must affect the composition of the flesh'.[15] Bovine tuberculosis fitted neatly into this interpretation especially once the tubercles

[10] John Burdon Sanderson, 'The Intimate Pathology of Contagion', *Twelfth Report of the Medical Officer of the Privy Council for 1869*, appendix xi (London, 1870), p. 229.

[11] See Hugh Scurfield, 'Measures for the Prevention of Tuberculous Infection by Milk and Meat', *Public Health* vii (1894/5), p. 106.

[12] 'Shall we eat Tuberculous Meat?', *BMJ* i (1890), p. 865.

[13] See 'Consumption of Diseased Meat', *Meat and Provisions Trades' Review*, 10 Nov. 1877, p. 443.

[14] *Tuberculous Meat: Proceedings at Trial under Petitions at the Instance of the Glasgow Local Authority against Hugh Couper and Charles Moore* (Glasgow, 1889), p. 170.

[15] *BMJ* i (1862), p. 348.

had become visible. Work on parasitic diseases also pointed to meat as a source of infection. This interest in the role meat played in causing tuberculosis was reinforced by growing anxiety about the transmission of other animal diseases through meat at a time when it was feared that there 'was hardly a cow-house . . . which was not continually furnishing diseased animals to the butcher'.[16] MOsH were also looking for a 'scientific' basis for destroying diseased meat whilst being uncertain about what meat was fit for human consumption and what was not. The idea that meat conveyed bovine tuberculosis was made easier to accept in part because research into outbreaks of food poisoning had already started to link meat with specific diseases, and studies had begun to claim that certain diseases (notably foot-and-mouth) were more virulent when fed to test animals.[17]

Elite veterinary surgeons, who were keen to assert their public health credentials and were already convinced of the zoonotic properties of the disease, were among the first to embrace the idea that meat from cattle with tuberculosis was dangerous. Leading veterinary manuals in the 1870s pointed to similarities between tuberculosis in humans and the condition in cattle and suggested that meat was an agent of infection.[18] Many drew on European studies to argue that 'man and other animals have become affected with the disease' where the 'flesh of consumptive subjects has been used as food'.[19] The council of the RCVS appointed a commission in 1880 to investigate the use of meat from diseased animals in order to provide the quantitative evidence that many believed was necessary. Tuberculosis was one of the twenty diseases covered by the investigation. The commission not only supported the idea that tuberculosis was contagious but also that meat played an important role in infection.[20]

Although veterinarians were quicker to accept the idea that 'the use of the flesh' from tuberculous animals 'should be carefully watched', claims that many doctors would 'not have imagined that it was necessary to condemn the carcases of animals affected with the disease on the grounds that there was a risk to the human subject of contracting the disease from the meat' proved unfounded. Doctors were more sceptical about the contagious nature of bovine tuberculosis than veterinary surgeons, but as studies began to highlight the role of 'juices and particles of the tainted animal' in conveying bovine

[16] *Annual Report of the Medical Officer of the Local Government Board for 1885* (London, 1886), p. xii.

[17] Anne Hardy, 'Food, Hygiene and the Laboratory: A Brief History of Food Poisoning in Britain, c.1850–1950', *SHM* xii (1999), pp. 293–311.

[18] See William Williams, *Principles and Practice of Veterinary Medicine* (Edinburgh, 1874), p. 342.

[19] R. Bradshaw, 'The Contagious Diseases (Animals) Act and Tuberculosis', *Veterinary Journal* ii (1881), p. 404.

[20] 'Inquiry into the Transmissibility of the Diseases of Animals to Man by Way of Flesh or Milk used as Food', *ibid.*, pp. 270–1.

tuberculosis attitudes began to change.[21] Because experimental evidence appeared to confirm that the disease was zoonotic, the belief that meat could transmit the disease offered a rationale for regulating the meat trade, as it provided empirical evidence to justify existing prejudices about the health consequences of eating diseased meat. A growing number of MOsH, already alarmed that 'the public are continually subjected to great risks through the unscrupulousness of some vendors of meat', were therefore convinced by these experimental studies and began to warn that tuberculous meat was dangerous.[22] For example, in 1877, William Hardwicke, MOH for Rotherham, in his evidence against a local butcher for the possession of tuberculous meat, was clear that 'the flesh of an animal suffering from Tuberculosis, Scrofula or Consumption, was unfit for human food.'[23] This was not an isolated case: three years later, the *BMJ*, reporting on a case before the Leicester magistrates, was heartened that a heavy fine had been imposed on a butcher selling meat 'infected with consumption'. The editor noted that the 'offence cannot be too seriously regarded' as 'experimental research leaves little reason to doubt that the consumption of tuberculous meat is extremely likely to produce . . . tuberculous disease'.[24]

By the mid 1880s, alarm about the widespread sale of meat from tuberculous livestock was being frequently expressed, and statements on tuberculosis regularly pointed to the dangers of meat. Robert Koch's identification of the causative agent for tuberculosis in 1882 helped confirm the role meat had in the transmission of the disease. Koch agreed with earlier researchers and pointed to 'the danger of flesh' of tuberculous animals.[25] Although his findings encouraged a new bacteriological framework to be adopted, many of the conclusions about meat as an agent of infection were the same. These views were reinforced by experimental studies that revealed that it was possible to transmit the disease from animal to animal by feeding them meat and especially organs from tuberculous cattle.[26] Many researchers were equally certain that the disease was introduced in food and spread through 'the digestive tract'. *The Lancet* therefore expressed 'no hesitation' in asserting in 1885 'that the carcase of an animal which has suffered from tuberculosis was unfit for human food'.[27] Evidence given by MOsH and veterinary surgeons in prosecutions for the sale of diseased meat reinforced the idea that tuberculous meat was unfit for food. For example, when the veterinary surgeon Edwin Faulkner was called to

[21] *Lancet* i (1880), p. 946.
[22] 'Tuberculosis', *Veterinary Journal* iii (1876), p. 474.
[23] 'The Flesh of Tuberculous Cattle as Food', *ibid.* v (1877), p. 68.
[24] *BMJ* i (1880), p. 99.
[25] *Berliner Klinische Wochenschrift* (Berlin, 1882), p. 230.
[26] Edward Klein and Heneage Gibbes, 'Experiments of Feeding Animals with Human and Bovine Tubercular Matter', *Annual Report of the Medical Officer of the Local Government Board for the Year 1884* (London, 1885), pp. 169–72.
[27] *Lancet* i (1885), p. 87.

give evidence against a butcher in Manchester found to be selling tuberculous meat, he could confidently claim that 'people who ate flesh in this state of disease were likely to catch the disease'.[28] Some were suspicious that the danger might be exaggerated, but even sceptics noted that the public should be shielded from eating tuberculous meat. As Arthur Newsholme, MOH for Brighton commented, 'the public have a right to be protected from every risk' and he went on to add that it was 'the duty of the MOH to seize the flesh of tuberculous animals'.[29] Those who were prepared to argue that tuberculous meat 'may be consumed with safety, however advanced' were increasingly dismissed 'as being reckless in the extreme'.[30]

If confirmation was needed the Privy Council's departmental committee on pleuro-pneumonia upheld the belief that meat could convey the tubercle bacillus. Although the committee dealt with probability and chance rather than proof of the injury, it not only accepted that a substantial amount of meat from cattle with tuberculosis was sold but was also dogmatic on transmission and the danger of meat. It wholeheartedly supported the notion that bovine tuberculosis was transmitted by 'swallowing, into the alimentary or digestive system'. The committee explained that 'the chance of the bacilli of tuberculosis being in the flesh or blood of animals affected by the disease was too probable to allow of the flesh of a tubercular animal being used for food under any circumstances'.[31] The same stance was adopted by the influential 1888 Paris congress on tuberculosis. Its delegates resolved that 'it is imperative that every possible means should be adopted . . . for the general application of the principle of seizure and general destruction in totality of all flesh belonging to tuberculous animals'.[32] *The Lancet* added that broad agreement now existed 'as to the transmissibility of tuberculous disease by means of food', a view endorsed in the following year by the International Veterinary Congress in Paris.[33]

Although dissenting voices continued to prompt unease, by the late 1880s most veterinary surgeons and an increasing number of doctors had been swayed by the 'numerous experiments on animals' that were felt to 'leave no doubt that the disease [bovine tuberculosis] can be contracted by tuberculous matter which has been eaten'.[34] As alarm at the amount of meat from cattle with tuberculosis being sold intensified, the belief that 'human phthisis frequently

[28] *Sanitary Record*, 15 Aug. 1889, p. 65; 'Bad Meat Cases', *ibid.*, p. 91.

[29] Arthur Newsholme, 'Condemnation of the Meat of Tuberculosis Animals', *Public Health* v (1892/3), pp. 4–5.

[30] George Reid, *Practical Sanitation: A Handbook for Sanitary Inspectors and Others Interested in Sanitation* (London, 1895), p. 253.

[31] *Departmental Committee appointed to inquire into Pleuro-pneumonia and Tuberculosis in the United Kingdom*, PP (1888) xxxii.

[32] *Congrés pour l'Étude de la Tuberculose chez l'Homme et chez les Animax* (Paris, 1889), p. 156.

[33] *Lancet* i (1890), p. 973.

[34] *Practitioner*, Sept. 1889, p. 226.

comes from the butcher's stall' gained ground. Claims that tuberculous meat had been 'eaten for years with impunity' were now dismissed as 'absurd' on the grounds that 'the evolution of a malady like tuberculosis is too slow to permit any causal relationship to be traced'.[35] Anxiety increased almost to a panic and *The Lancet* worried that 'it seems not unlikely that a greater amount of harm is done to public health by the ingestion of diseased meat than either the public or the profession have any idea of'.[36] As a result, MOsH and sanitary inspectors became more active in their efforts to protect the public from the sale of tuberculous meat. The *BMJ* echoed the view of many MOsH when it explained that the sooner tuberculous meat was thoroughly tackled 'the sooner shall we obtain a cleaner bill of health under the heading tuberculosis'.[37] Yet, despite the mounting alarm that came to surround tuberculous meat, the extent to which meat from cows with tuberculosis ought to be considered dangerous when the visible lesions were limited and the muscular tissue was apparently healthy, proved more divisive.

'What is diseased meat?'

If many by the 1880s could accept the idea that bovine tuberculosis was a zoonosis, and that meat was an agent of infection, the erratic sanitary practices and conflicting testimony in court highlighted the difficulty of determining the exact danger meat posed. As one MOH noted: 'meat which, after inspection, is pronounced fit for sale in one market' was 'liable to seizure in another because the inspecting authority happens to differ in opinion as to the extent of tuberculosis which may be dangerous'.[38] Even after Koch identified the tubercle bacillus and confirmed that the bovine and human forms of the disease were the same, the key question of how dangerous tuberculous meat actually was remained. Confusion surrounded the questions regarding whether the disease was localised or generalised, and the amount and distribution of the tubercles that justified a carcass being condemned as being unfit for human food. Here veterinarians and doctors broadly expressed differing views. What was at stake was more than a simple empirical question of the extent to which diseased meat was safe. Debate about localisation versus generalisation highlighted different ways of conceiving bovine tuberculosis. It also reflected different approaches to how bovine tuberculosis should be tackled, diseased meat condemned, and the public protected that were to frustrate attempts to regulate the meat trade.

[35] *MPC*, 23 Apr. 1890, p. 439.
[36] 'Unsound Meat', *Lancet* i (1885), pp. 87–8.
[37] 'Tuberculous Meat', *BMJ* i (1889), p. 1361.
[38] *Journal of Meat and Milk Hygiene* i (1911), p. 32.

In their understanding of bovine tuberculosis, most veterinarians adhered to a localised model of infection and supported the notion that infection was concentrated in the affected organs. This understanding of the disease reflected veterinary surgeons' concern with the 'health of the livestock economy' and an approach that stressed the value of practical observation. Here they drew on an organ-based pathology that had characterised medicine in the first half of the nineteenth century.[39] In addition, they made direct comparisons with other diseases and especially cowpox. They held that in diseased animals 'if there are no naked eye signs of tuberculosis in the carcass the condition of the viscera may be set aside, and this constitutes "localized" tuberculosis'.[40] William Williams, a stalwart of the veterinary community, was clear that if the tubercles were removed 'the flesh is perfectly good, fit for any man's table'.[41] The stance was confirmed by the work of European pathologists who suggested that tuberculosis was invariably limited to the organs. They argued that even when the bacillus was present in the muscular tissue, the numbers present were probably insufficient to transmit the disease.[42] Only in acute and very advanced cases, where the lesions and bacilli were widespread, was meat considered unfit for consumption. Koch agreed, pointing to the fact that bovine tuberculosis 'remains more or less localised'. This view was adopted by the 1885 International Veterinary Congress, which supported the need for 'partial seizure' of only the infected parts of a carcass.[43] In adopting this stance, serious and advanced cases could be seized and the public protected but the majority of meat from tuberculous cattle could be considered safe. George Fleming, the leading veterinary authority on bovine tuberculosis, noted that, although this was a relaxation of a 'sanitary point of view', it was essential in preventing 'serious economic sacrifice'.[44] By thinking of the disease in localised terms, veterinary surgeons sought to limit the role of sanitary officials involved in meat inspection to the identification and removal of obviously diseased organs. In doing so, they defended their position to determine the degree to which cattle were diseased, a position upon which they felt sanitary officials were encroaching.

Not all doctors were convinced by these views and this ensured that the nature of tuberculous meat and the extent of the danger to the public became a source of conflict between veterinary surgeons and MOsH. These doctors subscribed to a second school of thought which, in rejecting what they

[39] Michael Worboys, *Spreading Germs: Disease Theories and Medical Practice in Britain, 1865–1900* (Cambridge, 2000), pp. 60, 200.

[40] 'Tuberculous Meat', *Public Health* vii (1899/1900), p. 757.

[41] Williams, *Veterinary Medicine*, p. 346.

[42] 'The Use of the Flesh and Milk of Tuberculosis Animals', *JCPT* xii (1899), p. 244.

[43] 'Dangers of Consuming the Flesh and Milk of Tubercular Animals and the Means of Preventing them', *ibid.* i (1888), p. 266.

[44] George Fleming, *A Manual of Veterinary Sanitary Science and Police* (London, 1875), p. 393.

considered the 'nauseous doctrine of localisation', saw all meat from cattle with bovine tuberculosis as liable to convey the disease.[45] Such doctors saw diseased meat as a purely public health issue and defended their right to determine the 'fitness of flesh for the food of man'. As elite veterinarians became more forthright in their views on the relationship between animal diseases and humans, and tried to make inroads on public health, these doctors grew more vocal in their views.[46] They doubted whether the interpretation put forward by veterinary surgeons was a safe basis for protecting public health, arguing that veterinary surgeons were primarily concerned with defending agricultural interests and were not qualified on matters of human health. Many were sceptical of veterinarians' claims to expertise and actively resisted the encroachment of veterinary surgeons into human medicine, rightly interpreting their attempts to define tuberculous meat as a move to wrest control over meat inspection away from doctors (see chapter 8). Convinced that tuberculosis was a 'constitutional malady', they conceived of bovine tuberculosis in a different way that tended to exaggerate the danger of tuberculous meat. They argued that the localised signs of disease concealed 'a constitutional infection' and that the idea of localisation was based on a failure to distinguish between the disease and its manifestations.[47] With bovine tuberculosis hard to detect in cattle through a visual examination unless the condition was advanced, MOsH went on to argue that when the tubercles had become visible the disease had probably spread throughout the body. Many therefore recommended that 'the presence of even a trace of tuberculous deposit rendered the risk of partaking of any animal in which it appeared so great that the whole carcase should be at once condemned'.[48] Although this stance was later to be dismissed as too pessimistic, doctors remained sympathetic to a generalised model of infection, certain that it provided the safest approach for protecting the public. As the MOH for Preston noted, although condemning the whole carcass was 'somewhat wasteful', it was 'the safest and therefore the most satisfactory method of disposal'.[49]

Microscopy and laboratory evidence on how tuberculosis was spread through the body was used to justify the position adopted by MOsH. Ideas that rinderpest and other conditions such as smallpox were transmitted by blood had been proposed in the 1860s, whilst Koch's work on anthrax had also highlighted how the 'blood of tissue-juices' was an important medium in which 'bacilli

[45] Henry Armstrong, 'The Sale of the Flesh of Tuberculous Cattle for Human Food', *Veterinary Record*, 21 July 1900, p. 41.

[46] 'Unsound Meat', *Lancet* ii (1884), p. 1131.

[47] 'Important Trial Regarding Tuberculosis Carcases at Glasgow', *JCPT* ii (1889), p. 182.

[48] *Veterinary Journal* xxxv (1893), p. 304.

[49] *Annual Report of the Medical Officer of Health for the Year ending December 31st 1894* (Preston, 1895), p. 17.

multiply with enormous rapidity'.[50] The same framework was applied to bovine tuberculosis. Suggestions were made that tuberculosis was a blood disease; that the agent (after 1882 the bacillus) responsible could pass through the intestinal wall into the bloodstream.[51] Koch explained that the tubercle bacilli were seized upon by 'wandering cells' and carried away in the blood.[52] Edward Klein, 'the father of bacteriology' in Britain, agreed. He argued that because the bacilli were distributed through the blood 'no part of the animal in which even a single organ is visibly affected with tubercle can be held free of the tubercular virus'. For him this meant that there 'is a danger to the consumer of any part of the flesh of the animal'.[53] European studies supported Klein's claims and following this lead pathologists pointed to how 'there is nearly always a leakage of the bacillus into the lymphatics and the blood'. This made it easier to accept that the bacilli were present in all parts of the body hence rendering the entire carcass potentially dangerous.[54] John MacFadyean, who was emerging as a leading veterinary authority in the new science of bacteriology, went further. He declared that

> although tuberculosis may be strictly local to commence with there is the tendency, or there is the danger, at any rate, of it becoming general if the bacillus burst into the blood stream; and we can never declare with certainty that in any particular carcass this has not occurred.[55]

All those who subscribed to the view that the carcass should be seized when the tubercles were detected tacitly accepted the role blood played as the main carrier of the bacilli. Even critics acknowledged that it was probably an important 'method' of infection.[56] The Departmental Committee on pleuro-pneumonia confirmed this view. It warned that 'although bacilli may be found but rarely in the flesh, still the chance of their being present either there or in the blood is too probable to ever allow the flesh of a tubercular animal being used for food under any circumstances'.[57]

[50] 'Inoculation of Pulmonary Tubercles', *EMJ* xv (1869–70), p. 465; cited in Edward Klein, 'Infectious Pneumo-enteritis in the Pig', *Annual Report of the Medical Officer of the Local Government Board for 1877* (London, 1878), p. 208.

[51] Royal Institute of Public Health, *Transactions of the Congress held in Aberdeen* (London, 1901), p. 221.

[52] Cited in *Tuberculous Meat: Proceedings at Trial . . . Hugh Couper and Charles Moore*, p. 222.

[53] *BMJ* i (1889), p. 1249.

[54] *Practitioner*, Sept. 1889, p. 228.

[55] *BMJ* i (1889), p. 56.

[56] John MacFadyean, 'The Virulence of the Blood and Muscles in Tuberculosis', *JCPT* v (1892), pp. 22–30; Arthur Littlejohn, 'Meat as a Source of Infection in Tuberculosis', *Practitioner*, Jun. 1909.

[57] *Departmental Committee . . . into Pleuro-pneumonia and Tuberculosis.*

A generalised model of bovine tuberculosis was reflected in the literature directed at sanitary officials. In the *Food Inspector's Handbook*, Francis Vacher explained that the disease was 'not to be regarded as localized because it is only apparent in the lungs, for through the lungs circulates the whole blood, and sooner or late the bacillus is carried into the system'.[58] Newsholme, speaking at the 1892 Portsmouth conference of the Society of Medical Officers of Health (SMOH), dismissed the doctrine of localisation as fallacious. For Newsholme, it was 'important to condemn the carcase of an animal suffering from tuberculosis in the earlier active stage of the disease' because

> our general pathological knowledge shows us that traces of such disease may in some cases escape detection, while in others in which the complete absence of tubercles is secured, there may still remain the bacilli of tuberculosis or their spores, which have not yet produced tuberculous growths.[59]

Others warned that just because careful and prolonged microscopic examination or difficult bacteriological techniques had not revealed the presence of the 'tubercle spores' in meat this did not mean that they were not present and capable of developing into mature bacilli. Whereas this indicated an ongoing scepticism about the value of bacteriology in identifying disease agents, the generalised model allowed that meat could 'conceal the germ of phthisis' and should therefore be considered dangerous, thereby strengthening sanitarians' claims to regulate the meat trade.[60] This view was accepted by the annual meeting of the SMOH in 1892. It resolved that: 'the presence of tubercle, at any stage, in more than one part or organ of a carcase, or the presence of tubercle in any other than the primary stage' is 'sufficient and proper ground for the condemnation'.[61] As Henry Behrend concluded in an article in *Nineteenth Century*

> so grave a question as that of the transmissibility of tuberculosis from cattle to man by alimentation, it is certainly wiser to err on the side of credulity. Even if evidence adduced should be deemed insufficient to establish absolute certainty, it is certainly ample to warrant the adoption of stringent measures for the regulation of the sale of the flesh of animals thus affected.[62]

With the science uncertain, many MOsH felt it was best to be cautious.

[58] Francis Vacher, *The Food Inspector's Handbook* (London, 1893), p. 34.
[59] Newsholme, 'Condemnation', pp. 4–5.
[60] Thomas Walley, *A Practical Guide to Meat Inspection* (Edinburgh, 1890), p. 150.
[61] *Public Health* v (1892/3), p. 203.
[62] Henry Behrend, *Cattle Tuberculosis and Tuberculosis Meat* (London, 1893), p. 32.

Agricultural and trade reactions

Whilst the public was readily accused of 'fatalistic indifference' and were felt to be 'rather apathetic about consumption', this was not true of the meat and farming industry.[63] They became increasingly alarmed about bovine tuberculosis and the odium the disease directed at the meat trade. Structural changes in farming and the onset of agricultural depression in the 1870s made farmers and stockowners anxious about the implications of studies that asserted the zoonotic properties of bovine tuberculosis. Although the downward trend in agricultural prices was less marked in the livestock and meat markets, from the mid 1870s beef prices had started to fall.[64] Short-term price fluctuations did matter more to perceptions than long-term trends, but anything that threatened to depress prices by encouraging consumer unease created alarm among butchers and meat traders. Bovine tuberculosis appeared to do just that, especially as prices for dairy cows, the mainstay of the dead meat market, were endangered by the financial penalties enforced when diseased meat was seized, a practice which became increasingly common after the identification of meat as likely to convey bovine tuberculosis. As Clement Stephenson, a veterinarian in Newcastle opined, as more meat was seized farmers felt 'the pinch of tubercular disease, and they are taking more interest in it'.[65]

Meat traders and butchers initially agreed with the concerns about meat expressed by veterinary surgeons and doctors, seeing the sale of infected beef as bad for business. They called for harsher penalties for those selling tuberculous meat, and were anxious that the public did not consider the meat trade as having low morals. 'In the interests of trade' the *Meat and Provision Trades' Review* 'thoroughly condemn[ed]' the sale of tuberculous meat.[66] By the mid 1880s, the idea that bovine tuberculosis was communicable through meat was generally accepted in the agricultural press. For example, an article commissioned for the *Meat Trades' Journal and Cattle Salesman's Gazette* explained that 'experiments had proved' that the disease was transmissible from the animal to humans by the consumption of the 'flesh of the tuberculous animal'.[67] Butchers were advised on what to look for and were warned of the dangers of tuberculous meat. Feeling under threat, they wanted 'the confidence of the public' and hence took pains to demonstrate that they were aware of the danger posed by tuberculous meat, voluntarily surrendering diseased carcasses to sanitary authorities.[68]

[63] *Ibid.*, p. 20.

[64] E. H. Hunt and S. J. Pam, 'Prices and Structural Responses in English Agriculture, 1873–96', *EcHR* i (1997), p. 480.

[65] *Departmental Committee . . . into Pleuro–pneumonia and Tuberculosis*, minutes of evidence, p. 88.

[66] 'Penalties for Unsound Meat', *Meat and Provision Trades' Review*, 13 Sept. 1879, p. 371.

[67] MTJCSG, 37 Oct. 1888, p. 4.

[68] *Glasgow Herald*, 29 Apr. 1889, p. 13.

However, despite the support expressed for the idea that bovine tuberculosis was contagious, the meat trade was keen to defend a localised model of infection. For many butchers not 'a single case [has] been quoted where the health of any individual *has* suffered from the eating of animals only slightly affected with tuberculosis'.[69] For example, Henry Hill of the Sheffield Butchers' Association was certain that 'you may find tubercles in the wind-pipe, in the lungs, on the pleura, and on the diaphragm' but that 'there have been no signs of tuberculosis in the flesh'. For the Association, 'when the disease is merely local, and the infected parts are taken away, there is nothing injurious in the food meat which remains'. Butchers wholeheartedly agreed that meat 'obviously infected' was unfit for human consumption, but supported veterinarians' conception of a localised model of infection because it offered a means to protect themselves in a climate in which beef and live-meat prices had begun to fall steeply.[70]

As the number of prosecutions for the sale of tuberculous meat increased, and accusations were made of the meat trade's 'disgusting brutality' of selling 'flesh saturated with disease germs', butchers went on the defensive and pressed more forcibly for a localised model.[71] They felt threatened and were aware that too much stress on the dangers of tuberculous meat was bad for business. Butchers therefore acknowledged that bovine tuberculosis could be transmitted between animals and humans through meat, but disagreed violently with the generalised model adopted by many MOsH. As one meat inspector told the *Glasgow Herald*, by 1889 'there is a fear on the trade now'.[72]

'Perfect security against infection'

Whereas problems existed in determining the extent to which tuberculous meat was dangerous, some hope was found in the belief that cooking offered some protection. The suggestion that cooking or boiling acted as a purifier of food dated back to the eighteenth century. From the 1850s, it was argued that the same methods could also render diseased meat safe. This was used to downplay the threat of meat as a source of contagion. For Parkes, 'animal poisons' were 'neutralised or destroyed by the process of cooking', a view adopted by the Privy Council.[73] Studies of animal parasites (especially trichina spiralis), and the role of meat in their transmission, reaffirmed the value of

[69] *Ibid.*, 6 July 1889, p. 6.
[70] *Royal Commission Appointed to Inquire into the Effect of Food Derived from Tuberculous Animals on Human Health* (London, 1895), pp. 88, 93, 95.
[71] MPC, 26 Sept. 1883, p. 274.
[72] *Glasgow Herald*, 2 May 1889, p. 4.
[73] *Fifth Report of the Medical Officer of the Privy Council for 1862* (London, 1863), pp. 28–30.

cooking in providing 'perfect security against infection'.[74] The same benefits were ascribed to cooking meat from tuberculous animals. For example, Fleming noted in 1874 that 'cooking might be relied upon' to 'annul any pernicious properties'.[75] Others reached similar conclusions, explaining that cooking dramatically reduced the number of bacilli and hence rendered meat from cattle with bovine tuberculosis safe to eat. Butchers claimed that they had been arguing this for years, and staunchly defended the idea that cooking made diseased meat harmless in the hope that this would offset the bad reputation the meat trade was acquiring.[76] The utility of cooking was emphasised to explain why the incidence of bovine tuberculosis in humans was not higher. The problem of tuberculous meat could thus be transferred to the domestic sphere and defused by good domestic management.

Although it was agreed that 'the dangers of contracting tuberculosis from eating affected flesh would be practically *nil* if the meat was properly cooked' unease was expressed that this was often not the case.[77] Some contemporaries doubted the utility and effectiveness of cooking. The normal methods of 'boiling, roasting, or pickling', warned Arthur Whitelegge, the author of *Hygiene and Public Health*, 'affords but imperfect protection'.[78] This stance was supported by the Departmental Committee on pleuro-pneumonia, which asserted that 'the ordinary methods of cooking were often insufficient to destroy the bacilli buried in the interior of the limbs'.[79] Others argued that even if cooking destroyed the bacillus, it did not always 'destroy the virulence' of the bacillus, leaving the 'spores' and 'poison' intact.[80] These uncertainties echoed earlier debates about the role of disposition, seed and soil in the transmission of tuberculosis. There was evidence to support this view. For example, studies by James McCall, principal of the Glasgow Veterinary College, claimed that even after boiling for fifteen minutes, 'juice' from infected meat still caused tuberculosis in 35 per cent of their experiments with rabbits.[81] At best, it was felt that cooking reduced the ability of meat to produce tuberculosis. Although this might not prove a problem for the fit and healthy, even a few 'spores' or bacilli were believed to represent a danger to children and the weak as in the course of time it was felt that the spores could develop into mature bacilli.

[74] John J. Thudichum, 'Parasitic Diseases of Quadrupeds used for Foods', *Seventh Report of the Medical Officer of the Privy Council for 1864*, appendix vii (London, 1865), pp. 303–467; T. Spencer Cobbold, *The Parasites of Meat and Prepared Flesh Food* (London, 1884).

[75] George Fleming, 'The Transmissibility of Tuberculosis', *British and Foreign Medico-Chirurgical Review* liv (1874), p. 461.

[76] 'Tuberculised Meat', *Sanitary Record*, 16 Mar. 1891, p. 481; 'Sterilisation of Tuberculosis Milk and Meat', *JCPT* xiii (1900), p. 79.

[77] *Veterinary Journal* xxxv (1892), p. 305.

[78] B. Arthur Whitelegge, *Hygiene and Public Health* (London, 1894), p. 115.

[79] *Departmental Committee . . . into Pleuro-pneumonia and Tuberculosis*, p. xxii.

[80] *Veterinary Journal* xxx (1890), p. 4.

[81] 'Tuberculosis', *Veterinary Record*, 11 May 1895, p. 611.

Such a belief built on existing notions that tuberculosis spread more easily 'within and into weak bodies' and was used to explain why children appeared more susceptible to the disease.[82] Therefore, prolonged boiling or cooking was recommended, even though it was feared that the nutritional properties of the meat would be lost.

Although MOsH warned that 'bacteria are killed only when the cooking is very complete', there were concerns that, as Thomas Walley, the principal of Dick Veterinary College, put it, 'can proper cooking always be assured?'[83] Doctors expressed doubts about whether this did in fact occur. Here the dietary habits of the British, many of whom were believed to prefer their meat under-done or 'practically raw', were considered to pose particular problems.[84] A report for the Privy Council noted in 1864 that

> we may enter twenty cooks' shops, and we shall find the roast or boiled leg . . . which is being cut up red and underdone throughout its interior portion. The people like it, believing it more nourishing, as it is more juicy, and in a sense more tender.[85]

Under these conditions, Newsholme worried that MOsH were bound 'to assume that the flesh of any animal which we allow to pass may be eaten in a raw condition'. Consequently, he argued that 'cooking cannot be allowed to enter into consideration' when it came to limiting infection from bovine tuberculosis.[86] As one guide to sanitary inspectors explained in 1903, it was when cooking had not been complete that the danger occurred because 'this is always liable to happen; it is therefore desirable to prevent the flesh of animals that have had communicable diseases being used for food.'[87]

These concerns about cooking were carefully constructed within a class and gendered framework. Disquiet about food preparation among the urban poor had been articulated since the 1860s as part of efforts to reform working-class habits. Much of the attention was directed at women who were readily accused of being 'ignorant of all those habits of domestic economy' fitting for a good

[82] Michael Worboys, 'From Heredity to Infection: Tuberculosis, 1870–1890', in *Heredity and Infection: The History of Disease Transmission*, ed. Jean-Paul Gaudillière and Ilana Löwy (London, 2001).

[83] 'Cooking of Meat', *Public Health* x (1897/8), p. 237; Thomas Walley, *The Four Bovine Scourges: Pleuro-pneumonia, foot-and-mouth-disease, cattle plague, tubercle (scrofula)* (Edinburgh, 1879), p. 168.

[84] Jasper Anderson, 'Tuberculous Meat and its Exclusion from the Meat Market', *Public Health* iv (1891/2), p. 290; *Royal Commission Appointed to Inquire into the Administrative Proceedings for Controlling the Danger to Man through the use as Food of the Meat and Milk of Tuberculosis Animals*, part ii, minutes of evidence (London, 1898), p. 101.

[85] Thudichum, 'Parasitic Diseases', p. 410.

[86] Arthur Newsholme, 'On the Condemnation of Meat of Tuberculous Animals', *Transactions of the Sanitary Institute* xiii (1892), p. 308.

[87] Francis Allan, *Aids to Sanitary Science, for the Use of Candidates for Public Health Qualifications* (London, 1903), p. 158.

home.[88] However, notwithstanding a shift in how space and domestic technology were used in the home in the last decades of the nineteenth century, most working-class or poor families lacked cooking facilities. Overcrowding remained a problem into the Edwardian period, with many poor families forced to live in conditions in which 'rooms intended to be used as bedrooms or parlours have now to serve for the entire accommodation of a whole family'. Such an environment made cooking difficult or impossible, but even in many tenement homes 'any cooking was done on an open coal fire', which was expensive and inconvenient.[89] For most, 'the certainty of an economical stove or fireplace is out of the reach of the poor. They are often obliged to use old-fashioned and broken ranges and grates which devour coal with as little benefit as possible.'[90] The cost and inconvenience of cooking kept it to a minimum. Others relied on bakers. Looking back on her childhood in the 1860s, one woman remembered that 'no one had a cooking stove so the meat and some potatoes and a pudding were all put into one dish and taken to the bakehouse to be baked'.[91] Later evidence to the Scottish Royal Commission on Physical Training reported that a large number of 'housewives' had given up on cooking.[92] Many sanitary officials were all too aware of these conditions, encountering them on a regular basis as part of their sanitary work. It was no surprise, therefore, that in practical terms cooking was felt to offer inadequate protection when it came to tuberculous meat.

On the other hand, an insistence on 'wholesome if properly cooked' meat was reassuring, particularly to middle-class families at a time when the sale of diseased and tuberculous meat was considered ubiquitous. It linked measures to combat the dangers of tuberculous meat with efforts to reform the domestic habits of the urban poor, an approach that merged with campaigns to reduce tuberculosis by attacking the living and working conditions of the poor. Initiatives to provide the growing number of working-class housewives with domestic training and education in nutritional health were introduced. A range of organisations offered classes in domestic tasks and considerable advice was provided on cooking and on the purity of meat.[93] Such campaigns, although popular, failed to recognise the realities of many working-class homes,

[88] Edwin Chadwick, *Report on the Sanitary Condition of the Labouring Population of Great Britain* (London, 1842), p. 205.

[89] Derek J. Oddy, *From Plain Fare to Fusion Food: British Diet from the 1890s to the 1990s* (Woodbridge, 2003), p. 45; Anna Davin, 'Loaves and Fishes: Food in Poor Households in Late-Nineteenth Century London', *History Workshop Journal* xli (1996), pp. 168, 170.

[90] Maud Pember Reeves, *Round About a Pound a Week* (London, 1913), p. 59.

[91] Margaret Llewelyn Davies, ed., *Life as we have known it* (London, 1990), p. 14.

[92] Cited in David Smith and Malcolm Nicholson, 'Nutrition, Education, Ignorance and Income', in *The Science of Nutrition, 1840–1940*, ed. Harmke Kamminga and Andrew Cunningham (Amsterdam, 1995), p. 291.

[93] Elizabeth Roberts, *A Woman's Place: An Oral History of Working-Class Women, 1890–1940* (Oxford, 1984), p. 152; Joanna Bourke, 'Housewifery in Working-Class England,

but the assertion of the value of cooking bolted good domestic management onto the control of diseased meat by making it the personal responsibility of the good housewife. However, whilst evidence suggested that cooking did diminish the danger, it was felt that it was 'too uncertain to be trusted'.[94] Pointing to the deficiencies of cooking provided justification for preventive action and measures to curb the sale of diseased meat, something that sanitary officials and veterinarians increasingly sought.

By the 1880s, the sense that meat could transmit bovine tuberculosis had refocused debate about diseased meat onto the disease itself. Problems with identifying tuberculous livestock, even for experienced veterinarians, and doctors' efforts to define bovine tuberculosis as a public health issue, concentrated attention on protecting the consumer because, as the *Medical Press and Circular* explained, 'the consumer does not know that he is eating diseased meat [and] he cannot know that his subsequent sickness is caused by it'.[95] It was argued that because 'tuberculosis in one form or other carries off about one-sixth of the human race, so possible a source of infection as our meat supplies should be rigidly guarded'.[96] Given the 'imperfect security' offered by cooking, meat inspection was constructed as the main mechanism for protecting the public. Attempts to regulate the farming industry were therefore overlooked until the 1890s as to tackle the disease at an agricultural level was considered to have serious economic implications for the most dynamic section of the agricultural industry. However, the different models put forward by veterinary surgeons and doctors as to whether the disease was localised or generalised created a confusing picture that allowed considerable room for ambiguity and uncertainty. Opinions on the danger represented by tuberculous meat varied between regions, towns and individuals, and this had a large impact on local practices of meat seizure. Diversity in inspection and identification contributed to a climate in which disputes raged between whether the public would be better protected by condemning the entire carcass whenever a tuberculous lesion was discovered, or by removing only the visibly diseased parts. What emerged was a 'consensus of opinion as to the existence of such danger' from tuberculous meat but no 'unanimity as to its extent'.[97] One side of the debate suggested that no matter how small the risk, the public had a right to be protected; the other downplayed the danger. These tensions were highlighted in 1889 in Glasgow (see chapter 6), but all could agree by 1888

1860–1914', *Past and Present* cxliii (1994), pp. 183–5; W. Bowring, 'Address', *Journal of the Sanitary Institute* xv (1894), p. 414.

94 *Tuberculous Meat: Proceedings at Trial . . . Hugh Couper and Charles Moore*, p. 123.

95 *MPC*, 4 Apr. 1888, p. 357.

96 *Annual Report on the Health and Sanitary Condition of the Borough of Leicester during the Year 1889* (Leicester, 1890), p. 36.

97 *Veterinary Journal* xxxiv (1892), p. 73.

that bovine tuberculosis was 'one of the most important [issues] that at the present time claimed the attention of sanitarians' and that an adequate system of meat inspection was the best way of protecting the public.[98] How this system was to be organised and by whom proved difficult questions to resolve.

[98] *BMJ* ii (1888), p. 410.

5

Inspecting meat

How did debates about the nature of bovine tuberculosis and diseased meat translate into action? Once defined as a possible cause of sickness in the 1850s, efforts to identify and seize diseased meat were developed around a system of meat inspection that was part of a broader framework of nuisance control and fitted into local prescriptions for public health reform. The system of inspection established owed much to the tensions that shaped the management of public health and to the local officers appointed to remove the glaring sanitary deficiencies that characterised the Victorian urban environment. The appointment of a MOH often acted as a catalyst for this local sanitary reform and by the 1890s they were able to define an ideology of public health. However, it would be unwise to overstate development. The years after 1850 did see an increasing commitment to public health at local and central government levels but this was not a smooth or linear process. Inclusive and exclusive approaches existed side by side, and by the last decades of the nineteenth century the optimism of the early Victorian period had been replaced by a tension between the traditions of the public health movement and the search for the organisms that caused disease. The resulting uncertainty was reflected in the nature of meat inspection.

Just as the appointment of MOsH was often mediated by the 'grudging, almost incidental by-products of the intense manoeuvrings for advantage played out by the jealous interest-groups involved', meat inspection was hampered by similar constraints.[1] It was shaped by the nature of the local public health movement, the competence of sanitary officials, uncertainties about the nature of diseased meat, and by corporate interests. It was also subordinate to the need to deal with the multitude of sanitary problems facing Victorian towns, many of which so overwhelmed local authorities that they initially had little time for meat inspection. However, as bovine tuberculosis emerged as a public health problem in the 1870s and 1880s, meat inspection became an increasingly important local issue. Experimental evidence that the disease was contagious and transmissible to humans through meat not only redefined the problem of diseased meat but also increasingly came to underlie the rationale for meat inspection. In a climate in which it was believed that the public was 'utterly ignorant' as to what 'distinguished good from bad meat', it

[1] Simon Szreter, 'Healthy Government? Britain, c. 1850–1950', *Historical Journal* xxxiv (1991), p. 496.

was considered vital to defend them from the sale of diseased meat as 'a purchaser with a few pence in his pocket, and the cravings of hunger gnawing at his stomach, is not likely to exercise greater discrimination in the purchase of the necessaries of life'.[2]

However, by the late 1880s doubts were expressed about how well the public was being protected. The absence of a uniform approach to the seizure of tuberculous meat reflected the ongoing debate about whether bovine tuberculosis was localised in the organs or a generalised disease. Misgivings were voiced about the competence of those charged with inspection, casting suspicion on the effectiveness of meat inspection. Even so, the solution remained better meat inspection.

Meat and nuisances

The idea that diseased meat should be prevented from being sold was not new to the nineteenth century. Awareness that eating unhealthy or corrupt flesh could lead to sickness prompted measures in the thirteenth century to investigate the sale of diseased meat. In London, attempts to prevent the sale of bad meat were undertaken by the Honourable Company of Butchers. According to Blaisdell, the Company took its duties seriously, appointing wardens to carry out inspection as early as 1299.[3] Considerable emphasis was placed on appearance, and meat considered 'unfit for food' was seized and destroyed. Butchers' guilds in other cities equally strove to prevent members from selling diseased meat. Similar controls were introduced by other European states, but most supervision remained perfunctory until the late eighteenth century when efforts in several European states to improve the meat trade and establish a veterinarian sanitary police saw greater attention directed at preventing the sale of diseased meat.[4] An increase in recorded levels of epizootics and renewed interest in the possibility that some diseases (notably glanders, rabies and anthrax) could be spread from animals to humans, lay behind these efforts. Meat that was 'bad' and had a 'putrid tendency' was distinguished from meat that was injurious to health.[5]

[2] 'Meat Inspection', *Sanitary Record*, 15 Aug. 1890, p. 73; Thomas Walley, *A Practical Guide to Meat Inspection* (Edinburgh, 1890), p. 3.

[3] John D. Blaisdell, 'To the pillory for putrid poultry: Meat Hygiene and the Medieval London Butchers, Poulterers and Fishmongers' Companies', *Veterinary History* ix (1997), pp. 116–17.

[4] William Howarth, *Meat Inspection Problems, with Special Reference to the Development of Recent Years* (London, 1918), pp. 4–6.

[5] August Lydtin, George Fleming and M. van Hertsen, *The Influence of Heredity and Contagion on the Propagation of Tuberculosis and the Prevention of Injurious Effects from Consumption of the Flesh and Milk of Tuberculous Animals* (London, 1883), p. 38.

An awareness of the need to prevent the sale of unwholesome and (increasingly from the 1850s) diseased meat in Britain was part of this European trend. However, whereas contemporaries acknowledged that Britain led Europe in moves to lessen the worst effects of urbanisation on health, it lagged behind in measures to identify and seize diseased meat. At first, the Victorian prescription for public health left little room for meat inspection as it struggled to mitigate the worst consequences of unregulated urban growth. The sanitary problems resulting from overcrowding, inadequate housing, the accumulation of animal and human waste, contaminated water supplies and industrial pollution concentrated attention on removing environmental factors of obvious detrimental consequence and the sources of atmospheric stagnation, measures that were designed to prevent the spread of epidemic disease. At a practical level, therefore, it was often a desire to prevent the nuisances associated with the meat trade that initially shaped inspection.

By the mid-Victorian period, the meat trade had become a notorious source of nuisances. Cows were slaughtered in numerous small private slaughterhouses, many of which had grown up by accident and used buildings that were ill suited to the task. The blood and the stench of manure and offal, which attracted rats and polluted drains and water sources, were considered nuisances, prompting local sanitary authorities to use private and local acts to license and inspect slaughterhouses and use powers under the 1846–8 Nuisances Removal Acts to regulate offensive trades to control slaughterhouses. Action initially concentrated on the physical properties of the buildings and slaughtering practices.[6] These moves provided a focus for meat inspection.

This did not mean that the sale of diseased meat was completely ignored. Although there was no specific legislation, at least until the early twentieth century, the sale of diseased meat was dealt with under the existing system of nuisance inspection available to local authorities. The 1847 Town Improvements Clauses Consolidation Acts included sections that allowed medical officers to examine any 'animal, carcase, meat etc' intended or exposed for sale. Meat that was diseased could be seized. The Market and Fairs Clauses Act also made it an offence to sell unwholesome meat. Provision was extended by the permissive 1848 Public Health Act and by the 1855 and 1863 Nuisances Removal Acts. Both Acts had technical shortcomings, but clauses that permitted sanitary officials to seize diseased meat were part of what Hamlin has seen as a move by local bureaucrats to designate 'things, acts and conditions nuisances'.[7] Concern about food adulteration and provisions under the Adulteration Acts (1860, 1862 and 1872) also encouraged sanitary authorities to take a greater interest in the need to protect the poor from dangerous foods,

[6] J. Garrett, 'On Public and Private Slaughterhouses', *Public Health* vii (1894/5), p. 386.
[7] Christopher Hamlin, 'Public Sphere to Public Health: The Transformation of "Nuisance"', in *Medicine, Health and the Public Sphere in Britain, 1600–2000*, ed. Steve Sturdy (London, 2002), p. 196.

including diseased meat. The Acts established a more systematic inspection of markets and retailers.[8] However, definitions remained vague and a gap existed, as with much public health legislation, between the laws and their application.

London stood at the forefront of attempts to extend meat inspection. The appointment of MOsH in London and the passing of the 1855 Metropolitan Local Management Act created a network of sanitary officials whose duties included the regulation of slaughterhouses and markets. The problem of diseased meat was regularly discussed throughout the 1850s and 1860s by the MAMOH and by the Corporation of London. This encouraged MOsH in London to take an active interest in ending what they saw as a 'radically vicious' system that did not permit the 'examination of the health of animals killed, or the quality of meat'.[9] Henry Letheby, food analyst and MOH for the City, was a leading figure in these debates. He repeatedly argued that 'it is high time that some vigorous measures should be taken' to check the 'dangerous trade' in diseased meat. MOsH in London put pressure on the government to grant them greater powers in restricting the sale of diseased meat because they had encountered problems obtaining convictions when they had to prove that the meat was intended for human food.[10] However, the conflicting mass of unconsolidated public health legislation and divided responsibilities in London combined with doubts about the nature of diseased meat to frustrate efforts. Whereas Letheby and other metropolitan MOsH were active in seizing diseased meat at a parish level, it was only in 1877, following two decades of effort, that the Corporation appointed additional meat inspectors.[11]

Outside London, moves to increase the level of meat inspection were initially sporadic and uneven. The degree to which meat was inspected depended on the sanitary authority, the energy of the MOH, and the extent of local sanitary problems. The implementation of public health legislation was haphazard; legislation was permissive and local authorities could escape with doing comparatively little, although what was frequently interpreted as resistance to reform was in reality often bewilderment at the technical and legal complexities.[12] Marked variations existed between districts. Some authorities took their public health responsibilities seriously, others did not. The same was true of meat inspection. For example, in Merthyr Tydfil in 1849 diseased meat from Bristol was seized and, after examination, burnt in the market square, whilst in other towns little was done to regulate the meat

[8] Michael French and Jim Phillips, *Cheated Not Poisoned? Food Regulation in the United Kingdom, 1875–1938* (Manchester, 2000), pp. 33–4.

[9] St Giles Vestry, *Annual Report of the Medical Officer of Health* (1857), p. 34.

[10] *BMJ* i (1866), p. 33.

[11] MTG, Jan. 1867, p. 21; CLRO: Sanitary committee reports, 28 Apr. 1898.

[12] Christopher Hamlin, 'Muddling in Bumbledom: On the Enormity of Large Sanitary Improvements in four British Towns, 1855–85', *Victorian Studies* xxxii (1988/9), pp. 55–83.

supply.[13] As older MOsH retired their successors adopted stricter policies. Consequently, the number of prosecutions for the sale of diseased meat gradually rose and measures 'to prevent the sale of meat unfit for food' were implemented by a number of towns.[14] However, interest in the dangers of meat and the contagious properties of bovine tuberculosis was not initially matched by measures to combat the sale of tuberculous meat. Local authorities concentrated on addressing sanitary problems and this often left little time or energy for meat inspection.

A shift in inspection followed the 1875 Public Health Act. In consolidating and modifying existing legislation, it gave greater powers to sanitary authorities and provided the framework for public health administration until after the First World War. Food and meat inspection remained a minor concern; most of the 1875 Act was dominated by the pressing sanitary problems associated with water, sewerage and housing. However, the Act did extend measures to regulate the food industry and established mechanisms through which sanitary authorities could act, thus providing an impetus to meat inspection. This was reinforced by the interest generated by claims that bovine tuberculosis could be transmitted by ingestion of tuberculous matter. Rather than the Act being 'sparingly used', questions of meat inspection became more important to sanitary authorities as MOsH were increasingly asked to give their opinions on whether meat was diseased or not.[15]

By the mid 1880s, the examination of meat had become a regular feature of local public health work. A decline in mortality from the major infectious diseases (except tuberculosis) helped refocus attention, and a move away from ratepayer parsimony, together with the growth of civic consciousness and pride, created greater opportunities for action. Although the sanitary problems facing many towns ensured that inclusive policies were maintained, the search for the agents responsible for the major infectious diseases triggered greater attention on exclusive disease control. Moves to prevent the sale of tuberculous meat were part of this trend, with the extensive debates about the nature of diseased meat having a clear impact on local policy. Action was further encouraged by evidence of the unwholesome nature of the meat trade which made it clear that a large quantity of diseased and tuberculous meat was being sold and by studies that drew attention to the role played by meat in transmission of bovine tuberculosis. MOsH were further swayed by bacteriological studies that confirmed that humans 'have become affected with the disease'

[13] *Cardiff and Merthyr Guardian*, 26 May 1849, p. 3.

[14] For example, an inspector was appointed in Leeds in 1858 to prevent the sale of diseased meat; Lancaster town council suggested in 1869 that an officer be appointed to inspect slaughterhouses and markets: James Higgins, 'On Diseased Meat', *EVR* v (1863), pp. 670–1; 'Sale of Bad Meat', *BMJ* i (1869), p. 476.

[15] Peter J. Atkins, 'The Glasgow Case: Meat, Disease and Regulation, 1889–1924', *AHR* lii (2004), p. 164.

where the 'flesh of consumptive subjects has been used as food'.[16] This saw a rise in the number of prosecutions for the sale of tuberculous meat. Penalties were stiff, with fines ranging up to £20 and prison sentences of three months' hard labour.[17] However, effective controls relied on those local officials responsible for public health and those charged with inspecting meat. It was here that problems with the nature of meat inspection were seen.

'No special training or knowledge'

Although the second half of the nineteenth century witnessed significant changes in attitudes to public servants and an expansion of the role of the professional expert, meat inspection remained outside these developments. Precisely who was responsible and who was expert when it came to diseased meat were difficult issues to settle. Doctors, veterinary surgeons and butchers put forward competing claims of expertise. For doctors diseased meat was a public health problem and hence detecting diseased meat was the responsibility of the MOH. For the meat trade it was a matter that could only be settled by a competent butcher, whilst veterinarians claimed they possessed a body of knowledge that was superior to either group. Butchers characterised those responsible for inspecting meat as incompetent men 'who know nothing whatsoever of diseased conditions'.[18] Doctors and veterinary surgeons were equally disparaging of meat inspection, fearing that all too often it was undertaken by 'men whose general training has not in any way fitted them for the office'.[19] The diverse practices of inspection and seizure that occurred between sanitary authorities and the difficulties encountered in regulating the meat trade were commonly believed to be a direct result of poor, often inept inspection.

At the apex of the local system of meat inspection was the MOH. Although MOsH were first appointed in Leicester and Liverpool, London took the lead, with many provincial sanitary authorities slow to recognise the need for public health organisation. The 1872 Public Health Act made the appointment of MOsH compulsory and the MOH became an important member of local government. Dorothy Porter has argued that from this date a new professionalism emerged in public health practice, creating a national system of doctors who were responsible for the health of the community rather than the treatment of individuals. As a result, those holding the post changed from

[16] R. Bradshaw, 'The Contagious Diseases (Animals) Act and Tuberculosis', *Veterinary Journal* ii (1881), p. 404.

[17] *Meat and Provisions Trades' Review*, 13 Sept. 1879, p. 371.

[18] 'Who is to Decide What is Unfit for Food?', *ibid.*, 21 July 1877, p. 60; 'Meat Inspection', *Sanitary Record*, 15 Nov. 1890, p. 229.

[19] Thomas Walley, 'Meat Inspection', *JCPT* iii (1890), p. 112.

the elite medical men of the mid nineteenth century to the middle ranking doctors of the late Victorian period.[20] However, this process was not a smooth one: most MOsH were initially part-time, lacked tenure, and were poorly paid. There was no statutory qualification until 1888 and they initially occupied administrative rather than medical posts and were subject to considerable lay intervention. This had an important bearing on the duties they were expected (and able) to undertake. At a local level, much depended on the interests, energy and ability of the MOH, and this influenced the amount of diseased meat seized.

Under the 1848 Public Health Act all MOsH were expected to be qualified medical practitioners, but this did not mean that their medical training equipped them for their duties. This was clear when it came to meat inspection, justifying claims that many MOsH lacked the necessary experience to determine whether meat was diseased. In the ad hoc system of training that existed before the creation of the Diploma of Public Health (DPH) in 1871, instruction in what to look for in diseased carcasses was rudimentary. This reflected a situation whereby, according to *The Lancet*, most MOsH were 'not much better informed than other persons on questions relating to public health'.[21] The creation of the DPH not only formalised training but also, as part of an awareness of 'disease dissemination media', extended instruction in food safety.[22] The replacement of the sanitary ideal by the rise of state medicine did strengthen doctors' claims to expertise in all matters relating to public health, including meat inspection, but it was only with revisions to the DPH in 1900 to include the 'pathology of diseases of animals transmissible to man' that recognition was given to the role MOsH played in the identification of diseased meat.[23] Changes to the DPH highlighted the importance tuberculous meat had assumed in public health. However, for much of the nineteenth century the training directed at MOsH left them ill equipped to pronounce on diseased meat, even if they increasingly claimed the authority to do so. This led to problems with condemning carcasses.

The emphasis historians have placed on the importance of the MOH in local public health administration has ignored much of the day-to-day work of inspection. MOsH were assisted by a range of local functionaries who provided information and support which proved 'essential to the effective functioning of the local preventive authorities'.[24] Changes in the nature of

[20] Dorothy Porter, 'Stratification and its Discontents: Professionalisation and Conflict in the British Public Health Service, 1848–1914', in *A History of Public Health: Health that Mocks the Doctors' Rules*, ed. Elizabeth Fee and R. Acheson (Oxford, 1991), pp. 83–113.

[21] *Lancet* i (1872), p. 166.

[22] Dorothy E. Watkins, 'The English Revolution in Social Medicine, 1889–1911' (unpubl. PhD diss., London, 1984), p. 147.

[23] GMC, *Minutes* xxxvii, Nov. 1900, pp. 147–9.

[24] Anne Hardy, *The Epidemic Streets: Infectious Disease and the Rise of Preventive Medicine 1856–1900* (Oxford, 1993), p. 6.

public health as it moved away from the sanitary ideal broadened sanitary authorities' responsibilities. Between 1865 and 1870, the inspection of premises under such Acts as the Food Adulteration Act (1860), the Bakehouse Regulation Act (1865), and the Workshop Regulations Act (1867) became part of local sanitary activity. Over the next twenty years further duties were added, including the inspection of canal boats and infectious disease notification. In response, new duties were incorporated into the work of existing sanitary officials, and in some local authorities new posts were created. By the 1890s a patchwork of inspectors, which included those responsible for meat, had been appointed to carry out these new public health duties.

In this hierarchy, inspectors of nuisances (later called sanitary inspectors) occupied a central position. They were the main agent of the local authority in the 'lives of the mass of the people' and became the 'eyes, ears and nose' of the MOH.[25] The post had evolved from eighteenth century developments in police and turnpike administration and was given renewed importance in the Chadwickian sanitary ideal. Under the 1848 Public Health Act and the Nuisances Removal and Diseases Prevention Acts of 1848 and 1849, local authorities could appoint 'nuisance' inspectors. Although some towns had appointed nuisance inspectors before 1848, often employing police constables in the task, most were only spurred into action by the 1848 Act.[26] The responsibilities of the sanitary authority to inspect its district, rather than just responding to complaints, were gradually extended in the 1850s and 1860s, but it was the 1875 Public Health Act that required every urban and district council to appoint at least one full-time inspector.[27] Appointments often owed more to local perceptions of appearances and town pride than attempts to combat disease, and were frequently based on 'the vagaries of individual judgement and politics' rather than 'on a tangible assessment of ability'.[28] At first duties remained vague and the identification of a nuisance was often highly subjective, shaped by local concerns and complaints from neighbours. Although nuisance inspectors were dogged by anxieties about their low status, they emerged as the main 'instrument' of local public health work as their role gradually developed into positively seeking out defects in need of remedy.[29] Meat inspection was included as part of their duties and nuisance inspectors

[25] Janet Roebuck, *Urban Development in Nineteenth Century London: Lambeth, Battersea and Wandsworth, 1838–88* (London, 1979), p. 165; Albert Taylor, *The Sanitary Inspector's Handbook* (London, 1906), p. 17.

[26] This was true of Manchester and Cardiff: GRO: Catherine Bowler and Peter Brimblecombe, 'Control of Air Pollution in Manchester prior to the Public Health Act, 1875', *Environment and History* vi (2000), pp. 83–4; GRO: Cardiff local board of health, 1850–4, BC/LB/1.

[27] Anthony S. Wohl, *Endangered Lives: Public Health in Victorian Britain* (London, 1983), p. 195.

[28] Hamlin, 'Public Sphere to Public Health', p. 198.

[29] Roebuck, *Urban Development*, p. 54.

carried out the bulk of this inspection. Only in doubtful cases were they expected to report cases to the MOH for advice.

Initially, little practical sanitary knowledge was required for the post and there were no formal avenues for training or requirements for qualification until the 1880s. Instead, inspectors were expected to be 'men of fair education, able to write clear reports' and possess a 'practical knowledge of hygiene'. Individual character was important, and because it was felt that their work required more than 'mere knowledge', inspectors had to exercise 'tact and good common sense'.[30] According to Brimblecombe, this lack of requisite skills ensured that 'the office attracted men who had failed in other walks of life'. Roebuck paints an equally dismal picture. For her, nuisance inspectors in London were mainly unskilled workmen 'holding that which might almost be regarded as a sinecure office, an official recruited into the service of the vestry from the ranks of ex-sailors, ex-policemen, or army pensioners'.[31] In many towns too it was these same men who were responsible for inspecting meat. Many were part-time, paid low wages, had no security of tenure and were overburdened with duties; grievances that continued to animate inspectors throughout the second half of the nineteenth century as they sought to improve their professional status. This ensured that in the mid nineteenth century the majority of meat inspection rested with men untrained (or self-trained) in the signs associated with diseased carcasses.

Moves to make the appointment of nuisance inspectors compulsory saw demands for better training. In response, the LGB warned sanitary authorities against using police officers as nuisance inspectors. Although many took little notice and continued to use the police to identify nuisances and inspect meat, the recognition that standards needed to be raised prompted the Sanitary Institute to begin to offer examinations for surveyors and nuisance inspectors in 1877. The Institute had been founded the previous year and campaigned for the promotion and improvement of public health. It realised that to deal with the evils arising from insanitary conditions it was necessary to test the ability of all officers appointed to carry out sanitary work. A formal course of instruction for nuisance inspectors was started in 1885, which included classes on the identification of diseased meat. Essentially practical, the course was designed to develop the competence of inspectors and to offer local authorities a guide in appointments. Professional bodies and examinations provided an important role in creating a sense of identity and legitimacy. Through the courses offered by the Sanitary Institute experience was increasingly backed up by specialist technical training, a transition that was reflected in the change of title from inspector of nuisances to sanitary inspector.

30 'Sanitary Inspectors', *Sanitary Record*, 17 Oct. 1874, p. 273.

31 Peter Brimblecombe, 'The Emergence of the Sanitary Inspector in Victorian Britain', *Journal of the Royal Society for the Promotion of Health* cxxiii (2003), pp. 125–7; Roebuck, *Urban Development*, p. 78.

According to Roebuck, the creation of a qualification and the formation of professional bodies saw the office of nuisance inspector move from being 'a niche for unqualified amateurs' to one held by 'trained, experienced, expert local government officers'.[32] However, the extent of the transformation was not as great as Roebuck has claimed, especially since until the 1890s local authorities were under no obligation to appoint trained inspectors. In 1889, the *Sanitary Record* could still regard the tendency to eschew 'good practical' men in appointments as scandalous and having a detrimental effect on the quality of meat inspection. It worried that no inspector 'who is a policeman or a retired butcher, can acquire the necessary knowledge or should ever be trusted unaided to decide on the fitness or unfitness of meat for human food'.[33] Ongoing debates about expertise in identifying sanitary defects underlay concern as public health work shifted to include a greater emphasis on exclusive approaches, whilst fears about tuberculous meat emphasised the inadequacies of meat inspection. Despite attempts by sanitary inspectors to assert their professional identity, they continued to suffer from associations with corruption, whilst the SMOH remained sceptical of the examinations offered by the Institute. MOsH were afraid that sanitary inspectors might become cheap replacements for public health doctors; sanitary inspectors responded by expressing resentment at being the 'lackeys or the servants' of the MOH.[34] Even if sanitary inspectors acquired a level of independence, they were still essentially subordinate officers.

Although nuisance inspectors carried out the bulk of meat inspection, they were not the only local officers involved. Police officers and other local officials brought meat cases to the attention of the MOH. Concern about adulteration and food safety led to the appointment of market and food inspectors, adding another tier of officials to the local sanitary administration. Part of their work included the seizure of unwholesome or dangerous foods. As experimental studies drew attention to meat's role in the transmission of bovine tuberculosis, the need for these officers to identify tubercular carcasses and prevent their sale became increasingly important. However, as many market inspectors did not hold any qualifications, their authority was frequently questioned by butchers and they were deemed unable to provide the public with sufficient safeguards against the sale of diseased meat. In response, a number of local sanitary and market authorities started to appoint specialist meat inspectors.

When it came to the appointment of meat inspectors, those who dominated local authorities mainly sought to employ 'good practical' men with knowledge of the meat trade, elevating prudential knowledge above ontological expertise. In Liverpool, for example, all five meat inspectors were butchers by trade or

[32] *Ibid.*, p. 78.

[33] 'Meat Inspection', *Sanitary Record*, 15 May 1889, p. 528.

[34] Wellcome: SMOH council minutes, 5 Aug. 1891, SA/SMO/B.1/5; *Sanitary Inspectors' Journal* i (1895–6), pp. 260–1.

training, with the MOH convinced that this gave them a 'thoroughly practical experience'.[35] It was felt that 'a good butcher would know the tricks of the trade, and would be able to spot a great deal of diseased meat in that way'. This was clear in cases of bovine tuberculosis as, according to the chief inspector of meat for the City of London, 'lung disease is so common that every butcher knows it'.[36] Yet, despite the enthusiasm of some local authorities to appoint butchers, more often meat inspectors had little practical experience. Their occupational backgrounds were similar to those of early nuisance inspectors and it was suggested that many owed their positions 'more to favouritism than qualification'. Evidence to the second royal commission on tuberculosis provided an overview of the previous occupations of meat inspectors which included a

> plumber, carpenter, warehouseman and cheesemonger, sanitary inspector, butcher, vestry surveyor's clerk, ticket inspector, builder, plasterer, provision merchant, clerk in civil service, Inspector of Metropolitan Police, inspector of nuisances, stone-mason, carpenter, compositor, florist, surveyor, gasfitter, bricklayer and so on.[37]

For the meat trade, the type of men appointed clearly lacked the expertise necessary to identify diseased and tuberculous meat. This was perceived as the reason for the poor nature of meat inspection in many towns.

The ad hoc system of meat inspection created under legislation designed to prevent urban 'nuisances' was only as good as the officers charged with inspection. Although dominated by the MOH, most of the work was carried out by often overworked, poorly trained subordinates, many of whom possessed a limited knowledge of what to look for in slaughtered animals. It may well be suspected that this was often true of other public health posts. Growing alarm about bovine tuberculosis did see local authorities place greater emphasis on meat inspection and in some cases appointing specialist meat inspectors. However, the ongoing debate about expertise drew attention to problems with inspection.

[35] *Veterinary Record*, 28 June 1890, p. 695.

[36] *Royal Commission Appointed to Inquire into the Administrative Proceedings for Controlling the Danger to Man through the use as Food of the Meat and Milk of Tuberculosis Animals*, minutes of evidence (London, 1898), p. 72; W. Wylde, *The Inspection of Meat: A Guide and Instruction Book to Officers Supervising Contract-meat and to all Sanitary Inspectors* (London, 1890), p. 6.

[37] *Royal Commission . . . into . . . Controlling the Danger to Man through the use as Food of the Meat and Milk of Tuberculosis Animals*, minutes of evidence, p. 19.

'The most unpleasant and harassing duty'

The condemnation of diseased meat involved difficult questions of the sanctity of property and the identification of whether meat was diseased or not. Yet, despite growing concern about the effect on health of consuming diseased meat, at first inspectors and MOsH were given little advice. Most relied on Wilson's *Handbook of Hygiene* and a pragmatic approach.[38] It was only from the late 1870s that a growing body of literature was produced that advised sanitary officials on what to look for in diseased meat and the degree of infection that required the meat to be condemned. This was a response to increasing alarm about the role of meat as an agent of infection in bovine tuberculosis coupled to the extension of sanitary officials' responsibilities in relation to diseased meat under the 1875 Public Health Act.

Ideally, manuals explained, inspectors were to be present at the commencement of slaughtering. They were advised to watch butchers carefully as the 'butcher who desires to conceal or obliterate the presence of disease in a carcass has many tricks at his command' which careful observation would prevent.[39] The inspection of the whole carcass was recommended as inspectors had 'a much better chance of forming an opinion regarding the quality of the meat before it has been cut up'. Manuals described in detail the appearance of diseased meat (wet, pallid and sticky), and inspectors were told to make a systematic examination of the chest wall and organs, principally the lungs, liver, spleen and kidneys, as the most probable sites of tuberculosis. In supporting a generalised model of infection, most manuals called for seizure of all carcasses that displayed any signs of tuberculosis, a stance that reinforced the idea that the disease was a 'constitutional malady'.[40] However, inspectors were warned not to be too strict for fear that the 'meat supply would become seriously reduced' driving prices up 'beyond the reach of the poor'.[41]

The manner in which MOsH and their inspectors responded in practice when they encountered meat showing signs of bovine tuberculosis, or suspected of having the disease, varied widely. This was also true of other aspects of their work. The need to prioritise matters of importance at a local level and the emergence of new problems ensured that the everyday work of sanitary officials often differed greatly between towns. Inspectors were influenced by different conceptions of disease and this was true of meat inspection, especially given conflicting views about bovine tuberculosis as a localised or generalised disease.

[38] See George Wilson, *Handbook of Hygiene* (Edinburgh, 1873).

[39] William Robertson, *Meat and Food Inspection* (London, 1908), p. 58; Arthur Littlejohn, *Meat and its Inspection: A Practical Guide for Meat Inspectors, Students, and Medical Officers of Health* (London, 1911), p. 3.

[40] George Reid, *Practical Sanitation: A Handbook for Sanitary Inspectors and Others Interested in Sanitation* (London, 1895), p. 254.

[41] Taylor, *Handbook*, p. 298.

Inspection also depended on how rigorously sanitary officials undertook their duties. Although some employed elaborate strategies to seize diseased meat, the multitude of private slaughterhouses, markets and butchers ensured that most of those charged with meat inspection were forced to locate diseased meat as best they could. Inspectors were advised to meet 'cunning' with cunning, visiting markets or slaughterhouses suspected of dealing in diseased meat at 'night and early morning, when not expected' as it was mostly at these times that diseased livestock were slaughtered or sold.[42] As in many areas of public health work, information about the sale of diseased meat from vigilant neighbours and anonymous letters, and the co-operation of other municipal officers (especially the police) and communications between sanitary authorities were often vital. In some cases, butchers or slaughterhouses would be put under surveillance when nuisances were suspected.[43] However, manuals cautioned inspectors not to become 'the dupe of ill-will or idle rumour' spread by aggrieved neighbours or competitors, but to rely on their senses without being 'too prying or too much of the busybody'. Tact was often considered essential.[44]

Although sanitary officials encountered few difficulties in identifying putrid meat or meat treated with chemicals, the detection of diseased meat created considerable problems. Not only was meat the subject of a variety of 'frauds' to disguise its true nature, what constituted diseased meat remained difficult to quantify, especially when the telltale signs of disease were often hard to detect. The problem was particularly acute in cases of tuberculous meat. Although by the 1890s opinion was 'practically unanimous that in advanced stages of tuberculosis the consumption of the flesh should be prohibited' because 'the flesh was so deteriorated as to possess no longer the nature, quality, and properties' of wholesome meat, most tuberculous meat sold was not from cattle in the advanced stage of the disease.[45] This made it particularly difficult to detect. Bovine tuberculosis was protean in character: symptoms varied between animals, and cattle could suffer from the disease without showing any obvious signs of infection. In addition, it was frequently confused with other conditions (and in particular pleuro-pneumonia). Some veterinarians therefore considered that an accurate diagnosis 'even by the most expert clinical examiner' was always unreliable.[46] Contemporaries feared that this allowed substantial numbers of tuberculous beasts to escape detection, or equally led to beasts being wrongly condemned, strengthening demands for the appointment of inspectors with specialist training.

Inspection was further complicated by the habit of butchers of removing the obvious signs of disease. Butchers engaged in the sale of diseased meat went

[42] Wylde, *Inspection*, p. 98.

[43] CLRO: Sanitary committee, 25 June 1878.

[44] E. Smith, *Handbook for Inspectors of Nuisances* (London, 1873), pp. 29–31.

[45] *BMJ* ii (1892), p. 360.

[46] *Transactions of the British Congress on Tuberculosis* iv (1901), p. 111.

to 'immeasurable pains' to cover their tracks.[47] Even respectable butchers would remove the viscera or organs before inspection could take place, especially if they were diseased. This practice ensured that a large number of tuberculous animals were passed as sound, despite warnings that the removal of the pleura – the most common site for tubercles in cows – was a likely indication that the meat was diseased. As the *Veterinary Record* commented, 'when the carcases have once been dressed by a cunning butcher – as men who engage themselves in such a business usually are – it is very difficult indeed to detect anything amiss with them by even a moderately skilled person'.[48] It did not help that MOsH and their deputies were limited as to what they could do. Here the law protected meat traders and butchers. Carcasses were considered an article of commerce subject to depreciation. Sanitary officials were not therefore permitted to cut into the carcass and could not carry out extensive investigation without confiscation or a magistrates' order. For the most part, they therefore relied on a visual examination.

Even when the viscera were present, visual examination remained problematic. Despite claims by James Niven, the campaigning MOH for Manchester, that 'it is easy, after a little practice, for any intelligent trained man to examine carcases for the presence of tuberculous . . . glands', the identification of tuberculosis was not straightforward.[49] The pathological signs of the disease varied from organ to organ and were easily confused with other conditions. The poor state of medical and veterinary knowledge of bovine tuberculosis, its signs and symptoms, and whether it was a localised or generalised disease, complicated the situation and was a serious handicap to inspection. Given that meat inspection was only a small part of sanitary inspectors' duties until the 1880s, most were 'conversant with drainage, measurement, and examination of insanitary dwellings', but not meat inspection.[50] This made them unreliable protectors of the public when it came to diseased meat. Sanitary officers regularly admitted problems with determining if meat was tuberculous or not. Inexperience and a poor grasp of animal health and pathology saw cases of bovine tuberculosis wrongly identified and disagreements between officials as to the detection of the disease.[51] As a result, they would either condemn any meat suspected of coming from a tubercular cow or err in the favour of butchers.[52] Under these conditions, often only the most serious cases were condemned.

[47] Wylde, *Inspection*, p. 98.
[48] 'The Flesh of Diseased Cattle', *Veterinary Record*, 28 May 1892, p. 637.
[49] James Niven, 'Remarks on the Feasibility of Examining all Carcases for Tuberculosis', *Public Health* viii (1895/6), p. 82.
[50] Wylde, *Inspection*, p. 97.
[51] 'Unscientific Meat Inspectors', *Liverpool Courier*, 6 Dec. 1893.
[52] 'Meat Inspectors', *Veterinary Record*, 30 Sept. 1893, p. 185.

Looking back, Morrison Legge, secretary to the royal commission on tuberculosis of 1896 to 1898, explained that problems with identification had 'been the great stumbling-block hitherto in the way of dealing with the disease'.[53] A number of approaches were used to clear up doubts about whether meat was 'unfit for human consumption'. In some sanitary authorities, a butchers' jury was appointed, often with a veterinary surgeon as a member. Juries were supposed to protect the inspector as much as local butchers, condemning carcasses on the basis of a consensus. Often, however, they were trade moves to defend the interests of local butchers and ensured that only extensively diseased carcasses were condemned.[54] Most sanitary authorities therefore ultimately relied on local magistrates to determine whether meat was unfit. Yet, despite claims made by the meat trade, magistrates did not always side with inspectors. Butchers could be influential local figures. Magistrates in Sheffield, for example, were inclined to dismiss cases when the butcher concerned was a member of the local butchers' association, considering him respectable.[55] Inspectors regularly encountered difficulties in securing convictions. What constituted 'unwholesome' was easily challenged by local butchers giving evidence for the defence. As with other nuisances, a tension between public health and property rights caused problems. These were writ large when cases came to court, an arena in which medical expertise was subordinate to lay judicial authority. Here magistrates, who were rarely tutored in animal pathology, were required to settle often complex questions about whether meat showed signs of tuberculosis. A system that relied on local magistrates placed responsibility in the hands of those 'who have no very definite opinions of any kind on the subject of tuberculosis'.[56] Just as in cases of food adulteration, contradictory judgements frustrated prosecutions, creating a confusing picture and no clear precedents.

These problems are visibly demonstrated by a case in Bishop Auckland. When William Coulton, a butcher, was summoned for exposing meat for sale from an animal that 'had suffered, when alive, from tuberculosis', both sides employed expert witnesses. Their evidence highlighted the different conceptions of bovine tuberculosis and the competing expertise of doctors and veterinary surgeons in defining what meat was safe and what was not. Manson, the MOH for Bishop Auckland, was supported by Mulvey, a veterinary surgeon who argued that the carcass was 'studded with tuberculous deposits'. However, the defence called on another veterinary surgeon, John Farrow. Like many

[53] T. Morrison Legge and Harold Sessions, *Cattle Tuberculosis: A Practical Guide to the Farmer, Butcher, and Meat Inspector* (London, 1898), p. 19.
[54] 'Tuberculous Meat', *BMJ* ii (1889), p. 990; 'Tuberculosis in Meat', *Lancet* ii (1889), p. 1017.
[55] *Royal Commission . . . into . . . Controlling the Danger to Man through the use as Food of the Meat and Milk of Tuberculosis Animals*, minutes of evidence, p. 313.
[56] 'Tuberculous Meat', *BMJ* i (1889), p. 1359.

veterinarians, he adopted a localised model of infection and argued that the meat was from 'a perfectly healthy animal, notwithstanding the tubercles, which were of a hard bony nature, and entirely free from suppuration'. For Farrow, the meat was therefore fit for consumption, a view supported by several butchers and salesmen. The presiding magistrate concluded that the conflicting evidence suggested that 'there is doubt about the animal being unfit for human food', and on these grounds conviction was not justified.[57] Magistrates would regularly dismiss cases where the evidence conflicted, unable to decide which side was right. Under these conditions, local authorities and cautious sanitary officials were inclined to seize only the most obviously diseased meat.

'Slipshod methods' and 'casual arrangements'

Problems with identifying tuberculous meat and securing convictions were compounded by deficiencies in inspection that allowed a considerable amount of diseased meat to escape detection. Fears about bovine tuberculosis did increase the amount of diseased meat seized, but although it was recognised that inspectors made regular inspection of slaughterhouses and markets, reports of prosecutions for the sale of diseased meat suggested that 'slipshod methods of meat inspection [were] almost general in this country'.[58] Even in those areas in which attempts were made to improve inspection, the system was characterised as a passive and casual arrangement. Human and fiscal resources were limited. The administrative machinery for public health at a local level was often overburdened, limited by the structure of local government finance, which restricted the number of sanitary officials that could be appointed. Most sanitary departments were initially small. Many MOsH and sanitary inspectors were part-time, held a number of posts and were overworked. They often sandwiched meat inspection between other sanitary duties that were felt to be far more pressing in the struggle against epidemic disease.[59] Even in larger cities where a number of meat inspectors were appointed, as in London and Liverpool, the extent of the problem and the difficulties encountered ensured that it remained hamstrung by the nature of the meat industry.

The poor nature of meat inspection was hardly surprising given the amount of work expected. A typical example of an inspector's duties in regard to meat is that of Edinburgh's incumbent where he was required to

> be in attendance daily in the dead-meat market, to visit butchers' shops, be on the look-out at railway stations for carcases coming in from the country, call twice daily

[57] *Ibid.*, p. 25.

[58] *Departmental Committee appointed to inquire into Pleuro-pneumonia and Tuberculosis in the United Kingdom*, PP (1888) xxxii, pp. xiii–iv; Henry Behrend, *Cattle Tuberculosis and Tuberculous Meat* (London, 1893), p. 83.

[59] See GRO: Cardiff health and port sanitary committee, 9 Sept. 1890, BC/C/6/9.

> at the Police-office, and is likewise expected . . . to visit . . . byres, see if any sick animals are in them, and should he find any, to watch them narrowly, to ascertain how their carcases are disposed of.[60]

However, inspectors could not hope to be present in every slaughterhouse. In Bristol, there was only one meat inspector to cover 'some sixty' private slaughterhouses; in Hull the MOH had over 305 butchers' shops to inspect. Large markets, such as the live cattle market in Islington, also made inspection difficult and lessened the chance of 'being over looked'.[61] In these conditions, inspectors were often forced to deal with 'several thousands of cattle' in two or three hours.[62] Inspection on this scale was beyond many officers. It was widely recognised that inspectors 'could not see everything' and it was admitted that 'it was hardly possible to make a sufficient and satisfactory inspection'.[63] To cope with the workload, those charged with inspection often made only periodic visits. Nor were inspectors always on duty in the early mornings or at night when most diseased meat was sold. With slaughtering occurring at irregular times in multiple locations, it was impossible to control the meat trade. The best that could be hoped for, even by the most energetic inspector, was observation. As a result, it was felt that 'a considerable quantity of unsound meat is sold and consumed for human food' and only the most 'glaring[ly] unfit' meat was seized'.[64]

Inspection and the seizure of diseased meat were further frustrated by the low status, often tentative position of MOsH. Henry Armstrong, MOH for Newcastle, voiced a common complaint when he bemoaned the 'want of proper consideration for the office and officer'.[65] MOsH, despite having some local autonomy, were subject to considerable lay intervention. Sanitary officials faced complaints about the 'constant meddlesome surveillance and interference' and regularly encountered opposition.[66] As one contemporary wrote, the MOH 'If he does his duty' makes 'enemies of all the small property owners whilst he displeases the many whose first idea is to keep down rates'.[67] MOsH

[60] John Gamgee, 'Unwholesome Meat', MTG, 15 Dec. 1860, pp. 591–2.
[61] *Tuberculous Meat: Proceedings at Trial under Petitions at the Instance of the Glasgow Local Authority against Hugh Couper and Charles Moore* (Glasgow, 1889), p. 202; Headlam Greenhow, *Report on Murrain in Horned Cattle, the Public Sale of Diseased Animals, and the Effects of Consumption of their Flesh on Human Health* (London, 1857), p. 44.
[62] 'Tuberculous Meat', *BMJ* i (1889), p. 1359.
[63] John J. Thudichum, 'Parasitic Diseases of Quadrupeds used for Foods', *Seventh Report of the Medical Officer of the Privy Council for 1864*, appendix vii (London, 1865), p. 464.
[64] Garrett, 'On Public and Private Slaughterhouses', p. 388; *Report of the MOH of Newcastle-upon-Tyne* (1885), p. 42.
[65] *BMJ* i (1885), p. 181.
[66] Edward May, *Tenure of Office of Sanitary Inspectors* (Manchester, 1883), p. 6.
[67] Cited in John Davis, *Reforming London: The London Government Problem, 1855–1900* (Oxford, 1988), p. 88.

and sanitary inspectors were therefore often cautious about being too zealous for fear of dismissal. In urban areas, the strength of shopkeepers on public health committees created problems when it came to seizing diseased meat, whilst in rural districts farmers, who occupied a prominent position in local politics, were felt to be prejudiced against any sanitary improvement. Although this situation was not unique to MOsH, and applied to other local officials, the nature of urban local government and the strength of local farming and butchers' interests did hamper their efforts to regulate the meat trade, and when they seized diseased meat they did not always find support from the sanitary authorities that employed them. With more pressing sanitary problems facing local government, and given the wide range of areas MOsH were required to preside over, this ensured that until the 1880s diseased meat was often given a low priority. After all, it was a less pressing threat to public health than the problems presented by epidemics, sewerage, water and housing.

Opposition or co-operation?

At one level, butchers' and meat traders' associations were keen to stamp out the sale of diseased meat and aided local authorities in regulating the meat trade. Across the country, butchers voluntarily surrendered meat for destruction, aware that the consequences of being caught selling diseased meat were prosecution, a fine or imprisonment, all of which were bad for business. James Kay, a butcher in Middlesbrough, explained how 'as soon as they [butchers] saw something was wrong' they would go for the inspector.[68] Other butchers followed similar practices. In Crewe, local butchers were anxious to assist the MOH, aware that the seizure of diseased meat 'had a very bad effect on trade'. Similar levels of co-operation were reported in Sunderland, London and Leeds.[69] Medical officers were encouraged to persuade butchers to surrender voluntarily their diseased carcasses before they resorted to prosecution. They were advised to ask butchers and traders: 'Have you to your knowledge any unsound food upon your premises before I search them?' to give tradesmen the opportunity to co-operate.[70] Such co-operation was vital. Letheby explained that 'were it not that facility is offered by the salesmen for the detection of fraud his subordinates would be very much crippled in their operations'.[71] Far more carcasses were therefore destroyed by voluntary arrangement without resorting to a magistrate than by legal intervention.[72]

68 *Sanitary Record*, 25 Nov. 1893, p. 306.

69 *Veterinary Record*, 18 Dec. 1897, p. 353; *Report on the Health and Meteorology of Sunderland during the Year 1874* (Sunderland, 1874), p. 37; CLRO: Sanitary committee, 10 Oct. 1871; MTJCSG, 26 Jan. 1889, p. 9.

70 'Practical Meat Inspection', *Sanitary Record*, 23 Apr. 1908, p. 367.

71 'Unwholesome Meat', *Lancet* i (1867), p. 52.

72 For example, in Brighton only 10 cows were condemned by a magistrate between 1889

Not all butchers responded in this way though, highlighting how public health measures were often resisted by various interest groups. Although meat inspection did not encounter the same level of opposition as other areas of public health reform, notably the provision of sewerage or town water supplies, it remained a contentious issue. When Robinson took over as MOH for Greenock in 1874 and adopted a new policy of condemning all carcasses showing signs of tuberculosis, he met 'considerable' opposition. Francis Vacher encountered similar problems in Birkenhead.[73] Complaints were regularly made: in Merthyr, for example, butchers protested about the 'annoyance' to which they were subjected 'at the hands of certain officials'.[74] A system of local government that permitted public scrutiny and intervention by interest groups ensured that opposition could be effectively translated into moves to restrict efforts by inspectors to police the meat industry. This was the case in Brighton, where the powerful butchers' lobby resorted to intrigue to restrict the actions of Arthur Newsholme, the MOH.[75]

Most opposition was not as concerted as this, however. At a day-to-day level, butchers and the owners of slaughterhouses obstructed inspectors. Inspectors were offered bribes, and there is evidence to suggest that in some cases they accepted them.[76] More common were attempts to evade inspection; in some areas, lookouts were posted outside slaughterhouses to watch for inspectors.[77] Legal challenges were launched as to whether the meat was diseased, inspection had occurred at a 'reasonable' hour, the meat concerned had been exposed for sale, and over ownership of the meat seized.[78] In some cases, inspectors were assaulted verbally or physically in the course of their work. Randall in London noted that he regularly encountered meat salesmen who were 'exceedingly insulting', whilst in 1889 the Stockton police court heard how one butcher had 'threatened to fell' the sanitary inspector 'with a stick' when he tried to seize a tuberculous carcass.[79] Mainly, however, it was a question of trying to evade or cheat inspection.

and 1896, but a further 58 had been voluntarily slaughtered by butchers: *Royal Commission . . . into . . . Controlling the Danger to Man through the use as Food of the Meat and Milk of Tuberculosis Animals*, part ii, appendix F, p. 367.

[73] *Tuberculous Meat*, p. 73.

[74] *Merthyr Express*, 27 Feb. 1875, p. 15.

[75] John M. Eyler, *Sir Arthur Newsholme and State Medicine, 1885–1935* (Cambridge, 1997), pp. 69–71.

[76] 'City Commission of Sewers', *The Times*, 29 Nov. 1854, p. 5; CLRO: Sanitary committee reports, 19 Sept. 1871.

[77] G. Anderson, 'Slaughterhouses and Slaughterhouse Inspection', *Sanitary Inspectors' Journal* v (1899–1900), p. 80.

[78] Taylor, *Handbook*, pp. 295–7.

[79] CLRO: Sanitary committee, 27 Oct. 1885; 'Bad Meat Cases', *Sanitary Record*, 6 Dec. 1889, p. 303.

The main method of avoiding detection was through dressing the meat to hide or remove any signs of disease. Under these circumstances 'unless the inspector is more than usually alert and well qualified for his duty', as the veterinarian George Fleming explained, 'such a carcase may easily escape detection'.[80] In other cases, butchers would literally seek to hide evidence of disease. One case in Portsmouth illustrated how the butcher concerned had regularly cut up the diseased lungs 'into small pieces and buried them in a manure heap' to avoid detection. Others mixed good and bad meat together.[81] Inspectors regularly reported finding diseased meat that had been hidden in slaughterhouses and shops; of discovering trapdoors and hidden areas where diseased meat could be stored. Such practices were believed to be common.[82]

In areas where there was an active policy of inspecting meat, less scrupulous butchers would slaughter suspicious animals in outlying districts before bringing the dressed meat into town, often in the early hours. Here the chance of inspection was considerably lower. Under these conditions, 'such cases are only discovered by the merest chance'.[83] Attempts to extend meat inspection in Brighton, for example, saw Haywards Heath emerge as a market 'notorious for dealing in questionable cattle'.[84] Butchers would avoid towns where inspectors were particularly vigilant: when meat inspection was improved in Paisley 'a certain class' of butchers evaded 'the rigid inspection' by using the less well regulated Moore Street slaughterhouse in Glasgow.[85] As moves were made to prevent the sale of diseased meat, less scrupulous butchers took even greater efforts to frustrate inspection.

As the number of prosecutions increased and debate intensified about the threat from bovine tuberculosis, meat traders and butchers became less enamoured of meat inspection. Feeling their livelihoods threatened, they increasingly went on the defensive. From arguing that public health officials were 'protecting the butchers as well as the public', concerns began to be expressed that butchers were being persecuted and that the officers responsible for inspection were incompetent.[86] Many believed that they had legitimate grounds for complaint. As the MOH for West Hartlepool explained, when carcasses were condemned there was 'irritation on the part of the butchers, as the entire loss falls on them, and in most of the cases it seems to have been impossible to have detected the disease during life'.[87] Differences in approach

[80] George Fleming, 'The Transmissibility of Tuberculosis', *British and Foreign Medico-Chirurgical Review* liv (1874), p. 462.
[81] 'Tuberculous Beef', *Sanitary Record*, 16 Mar. 1891, p. 449.
[82] Walley, *Meat Inspection*, p. 6; 'Bad Meat Cases', *Sanitary Record*, 15 Sept. 1886, p. 137.
[83] A. Jasper Anderson, 'Public Slaughterhouses and Meat Inspection', *Public Health* vii (1894/5), p. 379.
[84] Eyler, *Newsholme*, p. 68.
[85] *Glasgow Herald*, 10 May 1889, p. 9.
[86] *Ibid.*, 21 Feb. 1880, p. 762.
[87] 'Compensation for Butchers', *Public Health* v (1893), p. 288.

to condemnation and disagreements over whether bovine tuberculosis was a localised or generalised disease created confusion and led to accusations of unfair practices. Local insurance or mutual aid schemes were started to protect butchers from the losses incurred from the increased likelihood that diseased meat would be seized. These were a reaction against moves by sanitary authorities to extend meat inspection and built on local traditions of mutual aid and trade associations. Wealthy farmers had established insurance schemes in the 1840s but outbreaks of rinderpest and pleuro-pneumonia had invariably forced them to close as they lacked the capital to cope with epizootics.[88] Bovine tuberculosis revived interest in insurance schemes, but many proved short-lived. Funds were often insufficient to cover losses from seizure, members were seldom sufficiently compensated, and farmers often refused to take part.[89]

At a national level, the National Federation of Butchers and Meat Traders' Associations was established in 1888 to defend butchers against the increase in prosecutions for the sale of tuberculous meat. Central to its campaigns was the demand for compensation, particularly in cases where there were no visible signs of tuberculosis when the animal was alive. Across Europe, compensation was considered key to measures to prevent the sale of diseased meat. In Britain, compensation had been introduced in sanitary legislation to prevent capricious prosecution but in the case of diseased animals it had been used to promote the slaughter of diseased cattle. Despite questions about the moral hazards of over-compensating the owners of diseased livestock, it had come to form an important component in policies to eradicate epizootics in cattle and swine, ensuring that slaughter and compensation had become firmly linked.[90] Butchers and the Federation did not want to go this far, but they did argue that as tuberculosis was hard to detect in cattle they should be compensated when they unknowingly bought and slaughtered tuberculous beasts. The medical and veterinary profession, aware that butchers were sometimes victims, expressed cautious support. However, compensation, and who should pay, became a stumbling block in debates on bovine tuberculosis.

Chaos

Despite protests by the meat trade about local efforts to improve detection and to prosecute butchers for selling tuberculous meat, by the late 1880s contemporaries were passing a dismal verdict on the nature of meat inspection. Aspirations to reform the meat trade and prevent the sale of diseased meat were not matched by achievements and words such as 'abominable' were

[88] John Gamgee, 'Cattle Diseases in Relation to Supply of Meat and Milk', *Fifth Report of the Medical Officer of the Privy Council for 1862*, appendix iv (London, 1863), pp. 206–7.
[89] *Select Committee on the Tuberculosis (Animals) Compensation Bill* (London, 1904), p. 31.
[90] *Hansard* clxxxi, 14 Feb. 1866, cols 488–92.

frequently used to describe meat inspection in many cities. Reformers lamented that the system was behind every European country and left to those with 'no special training or knowledge'.[91] Calls for the systematic inspection of meat were made with increasing frequency and conflicting claims were made over who was best suited to undertake inspection. All could agree, however, that better inspection and more extensive powers were required to protect the public from the sale of tuberculous meat.

Varying levels of vigilance were matched by the lack of uniformity in determining whether the meat in question should be condemned. Confusions over whether bovine tuberculosis was localised in the organs or a generalised disease made inspection difficult, and saw a wide range of seizure practices adopted depending on the views of individual inspectors and the attitudes of the sanitary authority. Wide differences in practice existed between newly trained officials and those who had been employed in the 1850s and 1860s, and between MOsH, their inspectors, and veterinarians. As *Public Heath* explained:

> in the inspection of tubercular disease in cattle we find one medical officer of health ordering the seizure of the whole carcase as unfit for food, even when the tubercular disease is confined to a single organ; another seizing the whole carcase when several organs are affected; and a third only seizing carcases in which the flesh is wasted and sodden, however extensive may be the tuberculous disease of the internal organs.[92]

The chaotic nature of meat inspection infuriated butchers. The need for uniformity in inspection became an important issue, reflecting wider concerns about the need for sanitary authorities and their inspectors to adopt uniform practices in relation to the practical methods involved in carrying out their duties. The uncertainties about what meat to seize and alarm about the threat from bovine tuberculosis came to a head in Glasgow in 1889.

[91] *Veterinary Journal* xxxi (1890), pp. 415–16.
[92] 'General and Local Tuberculosis', *Public Health* iv (1891/2), p. 205.

6

Making of an epoch?

By the late 1880s bovine tuberculosis was 'compelling general attention' as public debate about food safety increased in intensity and became a feature of Victorian reformers' concerns about poverty and health. The medical profession and lay public were alarmed about the alleged danger of transmission through eating infected meat and milk. Public health fears were fuelled by the press which widely reported cases of meat seized and carried articles that confirmed the risk to human health at a time when the tubercle bacillus was almost 'a topic of daily conversation'.[1] Apprehension also grew about the impact of the disease on the meat trade. Measures to control outbreaks of animal disease and the relative absence of epizootics only served to highlight the losses from bovine tuberculosis. Local authorities had also become more active in their efforts to regulate the meat trade: the number of trials for the sale of diseased meat had risen sharply as awareness of the potential threat from bovine tuberculosis grew and greater interest was expressed in food safety.

However, a sense of uncertainty remained. Disagreement continued around whether the disease was localised or generalised and hence the grounds on which meat from tuberculous livestock should be condemned. This doubt encouraged the Privy Council to extend the remit of a Departmental Committee established in 1888 to inquire into pleuro-pneumonia to include bovine tuberculosis. It confirmed the danger from bovine tuberculosis and proposed a combination of prevention and eradication measures.[2] A 'congress for the study of tuberculosis in man and animals' was held in Paris shortly after the committee reported under the presidency of the French veterinarian Jean-Baptiste Chauveau. He had asserted the communicability of tuberculosis through meat and delegates to the conference reinforced his views, recommending Europe-wide steps to 'obviate the dangers of tubercular infection by alimentation'.[3] The congress drew public and professional attention to the unsatisfactory state of meat inspection in Britain.

It was against this background that attempts were made in Glasgow to prosecute a butcher and salesman for selling tuberculous meat. The resulting 'dead meat drama' overshadowed the report of the Departmental Committee

[1] Edgar Crookshank, 'The Tubercle Bacillus', *JRAS* (1891), p. 83.

[2] *Departmental Committee appointed to inquire into Pleuro-pneumonia and Tuberculosis in the United Kingdom*, PP (1888) xxxii.

[3] 'Congress for the Study of Tuberculosis', *JCPT* i (1888), pp. 262–75.

and the congress, focusing national debate and underscoring disagreement over the threat from bovine tuberculosis. The resulting furore saw the appointment of a royal commission. However, rather than being 'epoch making' and 'momentous in the extreme', the Glasgow trial epitomised the growing unease about the risk from tuberculous meat. The trial was not, as Atkins has suggested, symptomatic of the early sparring between 'a number of health-conscious local authorities' and a 'profit-oriented butcher trade' that 'paid little or no attention to disease'.[4] Nor did it see the conversion of veterinarians to germ theories.[5] Many veterinarians had already accepted a contagious model of bovine tuberculosis. Rather, the trial reflected the increasing efforts by local authorities to prosecute those caught selling carcasses displaying signs of tuberculosis and attempts to determine whether they could condemn a whole carcass, 'however well it may appear', if there was any evidence of the tubercle.[6]

The issues brought to light by the trial, and the resulting royal commission, saw an attempt to define a clearer understanding of bovine tuberculosis and reform meat inspection. Although Vernon has claimed that by the turn of the century 'there was no convincing evidence that tuberculous cows could transmit the disease to humans', this is precisely what the royal commission sought to provide.[7] It helped fashion a British model that bovine tuberculosis was a threat to public health even if the extent of the danger remained uncertain. Although it took a second commission (1896–8) to outline recommendations on how tuberculous meat should be tackled, the first commission not only provided a new taxonomy but also spurred local efforts to regulate the meat trade.

The Glasgow case

By the 1880s Glasgow had become an important centre for the importation and distribution of cattle in the region. Its role in these trades had expanded rapidly under the encouragement of the town council, which opened a new dead meat market in 1879. However, efforts in the mid 1850s to improve the meat trade in Edinburgh also saw Glasgow emerge as a centre for the disposal of diseased meat in Scotland. This was facilitated by the poor system of meat inspection that existed in the city. Although provisions under the 1847 Markets and Fairs Clauses Act included measures to prevent the exposure of

[4] *MTJCSG*, 1 June 1889, p. 12. Peter J. Atkins, 'The Glasgow Case: Meat, Disease and Regulation, 1889–1924', *AHR* lii (2004), p. 162.

[5] Michael Worboys, '"Killing and curing": Veterinarians, Medicine and Germs in Britain, 1860–1900', *Veterinary History* vii (1992), pp. 53–71.

[6] The term 'dead meat drama' was first used by the *Veterinary Record* to describe events in Glasgow: *Veterinary Record* i (1888/9), p. 307.

[7] Keith Vernon, 'Pus, Sewage, Beer and Milk: Microbiology in Britain, 1870–1940', *History of Science* xxviii (1990), p. 301.

'unwholesome provisions', with powers over meat extended by the 1865 Glasgow Markets and Slaughterhouses Act and the 1866 Glasgow Police Act, the sanitary problems facing Glasgow ensured that moves to secure sanitary reform absorbed the energies of the police department's Health Committee.[8] Despite the appointment of an inspector of police to inspect markets, it was only following pressure from the Glasgow United Fleshers' Society in 1877 that an additional constable was employed to inspect 'every carcase, or part of a carcase' offered for sale as food in an attempt to improve the local meat trade. However, the constable was overworked and the Health Committee's two food inspectors were 'quite unequal to the proper discharge of the duties imposed upon them'.[9] In addition, it was believed that local butchers had entered into a 'bargain' with the magistrates, ensuring that only the most severely diseased carcasses were condemned.[10]

As national debate about the danger of meat intensified, the system of meat inspection in Glasgow was increasingly perceived as inadequate. Concerns about the public health danger of diseased meat were first voiced in Glasgow in 1882 after suspicions were raised that cattle from a dairy company in Renfrew suffering from an 'outbreak of virulent disease' were being sold at the dead meat market. On investigation, James Russell, the MOH, assisted by James McCall, principal of the Glasgow Veterinary College, confirmed that the meat 'was likely to pass on the disease' to humans.[11] Russell was more active in his duties than his predecessor. Aware of European studies on bovine tuberculosis, he was convinced that meat could transmit infection. Russell was anxious to improve the meat trade given the considerable amounts of diseased meat entering the city. He felt that the existing system of meat inspection was inadequate to protect the public and therefore suggested that responsibility for inspection should be transferred from the Markets Commission to the Health Committee.[12] In doing so, he hoped to bring meat inspection under his control and extend his authority over the city's public health infrastructure. However, the strength of the local meat trade on the town council saw to it that no action was taken and successive efforts by Russell to improve inspection were frustrated.[13]

Despite growing concern about the nature of the meat trade in Glasgow, it was not until 1888 that McCall and Russell were asked to investigate measures for 'the protection of the public against the danger of their eating the flesh' of

[8] A. K. Chalmers, ed., *Public Health Administration in Glasgow: A Memorial Volume of the Writings of James Burn Russell* (Glasgow, 1905), p. 28; Brenda White, 'Medical Police: Politics and Police', *Medical History* xxvii (1983), pp. 407–22.

[9] GCA: Special subcommittee on diseased meat, 14 Feb. 1882, MP9.24; Russell to Lang, 26 Oct. 1885, MP20.598.

[10] 'Inspection of Meat', *JCPT* ii (1889), p. 142.

[11] GCA: Committee of health, 23 Jan. 1882, E1 20/6.

[12] GCA: Joint committee on the inspection of dead meat, 30 Jan. 1891, MP29.170.

[13] GCA: Russell to Lang, 26 Oct. 1885, MP20.598.

tuberculous livestock. Growing alarm about bovine tuberculosis had seen Paisley sanitary authority take the initiative to deal with the disease following a bad outbreak in the town in 1887. It encouraged other local authorities to follow suit.[14] With activities in Paisley attracting publicity, concern was expressed that 'there are few towns in the three kingdoms, if any, where there are so many tuberculous and emaciated animals sold in the open market as in Glasgow'. McCall believed that this could be explained by inadequate meat inspection, which ensured that it paid the 'persons who traffic' in diseased meat 'to feed the inhabitants of Glasgow and the West of Scotland on the abominable carrion'.[15] He and Russell, in line with other public health officials, concluded that the only solution was for more meat inspectors to be appointed to ensure better surveillance. As in other cities, faith was placed in the value of inspection as part of an accepted approach to nuisance removal. Under pressure, the Markets Commission agreed that the appointment of additional inspectors was 'urgently required'.[16]

The *Glasgow Herald*, aware of these debates and mounting public concern, announced its intention in April 1889 to conduct an investigation into the local meat trade. It revealed that the city's meat market in Moore Street was a 'receptacle for doubtful and diseased meat from every part of Scotland and many parts of Ireland'. The *Herald* claimed that the standard of inspection in the markets was so low that meat accepted there would be condemned elsewhere, a situation the paper considered a scandal when there was 'practical unanimity among scientific, medical and veterinary experts that a tuberculous animal is dangerous'.[17] Throughout April, the *Herald* published more and more evidence that highlighted the shocking nature of the local meat trade and the poor system of inspection.[18] Local concerns reflected national debate. The more respectable members of the trade joined the clamour for reform and decried the 'cheap and nasty' meat being sold in Glasgow, worried about the damaging effects on trade as local sales of cattle fell.[19] The publicity generated by the *Glasgow Herald* encouraged the Health Committee to increase its efforts to prevent the sale of tuberculous meat as part of Russell's campaign against tuberculosis. It was conscious that meat inspection was of a 'perfunctory character' and was keen to take over responsibility from the Markets Commission, which, because it was dominated by local trading interests, was reluctant to act. In a bid to improve inspection, provisions under local Police Acts were abandoned in favour of more extensive powers available under the 1875 Public

[14] 'Meat Supply of Glasgow', *BMJ* i (1886), p. 710; GCA: Joint committee on the inspection of dead meat, 30 Jan. 1891, MP29.170; Committee of health, 5 Dec. 1887, E1/20/11.

[15] 'Tuberculosis in Cattle', *BMJ* i (1888), p. 713.

[16] GCA: Joint committee on the inspection of dead meat, 30 Jan. 1891, MP29.170.

[17] *Glasgow Herald*, 12 Apr. 1889, p. 10; *ibid.*, 20 Apr. 1889, pp. 4, 7.

[18] *Ibid.*, 24 Apr. 1889, p. 9; *ibid.*, 25 Apr. 1889, p. 4; *ibid.*, 29 Apr. 1889, p. 13; *ibid.*, 7 May 1889, p. 4.

[19] 'Diseased Meat in Glasgow', *MTJCSG*, 27 Apr. 1889, p. 9; *ibid.*, 4 May 1889, pp. 6, 12.

Health Act. As a result, greater authority for inspection was transferred to Russell.[20]

Whilst these efforts mirrored attempts by other authorities to regulate the meat trade in response to growing fears about bovine tuberculosis, it was the Health Committee's decision to test its right to prevent the sale of tuberculous meat that saw events in Glasgow attract national attention. The *Glasgow Herald* and the local meat trade had long been anxious for a test case to clear up confusion over what meat should be condemned.[21] The issue was pushed by the seizure of two carcasses affected with tuberculosis at the Moore Street slaughterhouse by Peter Fyfe, the chief sanitary inspector. Common to many nuisance inspectors, Fyfe had little experience in condemning meat and hence called in Russell, McCall and William Young, the police surgeon, to determine the extent to which the meat showed signs of tuberculosis. Russell and McCall supported germ theories of disease and role meat played in infection and declared the meat unfit. However, Young worked from a different perspective and could find no evidence of tubercles. Unlike his colleagues, he did not 'believe in the bacillus tuberculosis'.[22] It was here that problems started. An attempt was made to get Hugh Couper and Charles Moore, the butcher and salesman deemed responsible, to voluntarily destroy the carcasses. However, when they refused because of the conflicting opinions expressed about the meat the case came to court.[23]

The seizure of the two carcasses reflected a change in the nature of meat inspection in Glasgow. Police inspectors with limited knowledge of animal disease had traditionally passed meat 'if the flesh looks good' and required only the obviously diseased parts to be removed.[24] However, in 1889 there was a shift in emphasis as inspection was placed under Russell's control. Like many MOsH he accepted the idea that bovine tuberculosis was a generalised infection. He agreed with the Departmental Committee on pleuro-pneumonia and tuberculosis that 'the chance of the bacilli of tuberculosis being in the flesh or blood of animals affected by the disease was too probable to allow of the flesh of a tubercular animal being used for food under any circumstances.' He therefore felt that no matter how localised the tubercles the entire carcass had to be condemned, a policy he actively pursued. When Russell assumed responsibility for meat inspection, it was his views that prevailed and inspectors were instructed to 'pass nothing [they] could see a speck of disease upon'.[25] This marked a shift in inspection practices and saw the Health Committee

[20] GCA: Joint committee on the inspection of dead meat, 30 Jan. 1891, MP29.173.
[21] *Glasgow Herald*, 7 May 1889, p. 4.
[22] 'Important Trial Regarding Tuberculosis Carcases at Glasgow', *JCPT* ii (1889), p. 182.
[23] *Tuberculous Meat: Proceedings at Trial under Petitions at the Instance of the Glasgow Local Authority against Hugh Couper and Charles Moore* (Glasgow, 1889), p. 5.
[24] GCA: Tuberculosis in cattle, 4 Jan. 1888, MP17.301; *Glasgow Herald*, 22 Apr. 1889, p. 7.
[25] 'Sale of Diseased Meat in Glasgow', *MTJCSG*, 4 May 1889, p. 7.

take steps to prosecute Couper and Moore. Whether this was a move to restore public confidence in the meat trade following the *Herald*'s campaign, or to test the committee's new responsibility, is uncertain.

From the start it was widely acknowledged that the prosecution of Couper and Moore was a test case, not only of the right to condemn tuberculous meat but also on what basis that condemnation should be made. It was a combination of these factors, combined with local publicity, that helped make the Glasgow trial a *cause célèbre* just as a series of official and professional pronouncements appeared on bovine tuberculosis. During the trial local concerns about the meat trade fused with national interest in bovine tuberculosis to demonstrate the conflicting views of the dangers of meat.

Cause célèbre

The trial under Sheriff-Principal Berry lasted four days. As in many meat cases, the stances adopted were contradictory, full of hyperbole. Evidence quickly revealed marked differences between MOsH, veterinarians and butchers on the degree of danger from tuberculous meat. At the same time it showed how ideas about the disease had shifted since the 1870s, something the lawyers often found difficult to grasp. Whereas there was general agreement that the carcasses showed evidence of inflammation, contradictory opinions were expressed about whether they were tuberculous and if the meat was unfit for consumption. The key issue was the question of localisation and the extent to which meat from tuberculous livestock was dangerous to health. These issues had a resonance beyond Glasgow and raised fundamental questions about the nature of meat inspection and bovine tuberculosis. Although other trials had raised similar concerns this was the first major court case following the reports of the Departmental Committee and Paris congress. What made the trial in Glasgow more dramatic was that both sides employed 'the most eminent scientific evidence' in the form of pathological studies.[26] Despite the growing value ascribed to experimental studies in defining disease, these studies only demonstrated the contradictory views about the dangers of tuberculous meat. At certain points, the trial appeared to hinge on conflicting experimental evidence and provided a forum in which disagreements about what tuberculous meat was were publicly aired.

Witnesses for the prosecution were national advocates of the dangers of bovine tuberculosis and campaigners for better meat inspection, many of whom had given evidence to the Departmental Committee the year before. They drew heavily on the findings of the 1888 Paris congress, and the prosecution tried to claim a more scientific grasp of the subject. Like Russell, they stressed a generalised model of infection, asserting the role played by blood and its

[26] *BMJ* i (1889), p. 1131.

circulation in the transmission of the bacilli. They uniformly argued that localised tubercles warranted total condemnation to protect the public from infection because, as one witness explained, 'we can never declare with absolute certainty' that the bacilli had not 'burst into the blood stream'. Some even staked their reputations on it. However, when questioned, the expert witnesses became vague: they were frequently forced to admit that their evidence was limited; that the science was uncertain. As Arthur Littlejohn, former tutor at the RVC, admitted the science was 'guess work' but that this did not matter as 'if an animal is diseased it is unfit for human food, and common sense rebels against it being used as human food'.[27]

Experts for the defence did not always come across well and few were considered leaders in their field. They assigned a different value to experimental studies and presented counter arguments that bovine tuberculosis was a localised infection, going on to argue that meat from diseased livestock played only a limited role in the transmission of the disease when compared with inhalation of the bacillus. Witnesses for the defence maintained that the two carcasses were fit for human consumption and explained that they would have eaten the meat as there was 'nothing whatever' to suggest it was 'infective'.[28]

The two stances emphasised the different interpretations of tuberculosis and the nature of infection. Summing up, Jameson for the defence claimed that no evidence had been presented that eating meat from tuberculous cows was dangerous and suggested that a 'mountain' was being made out of a 'tubercle'. Comin Thomson for the prosecution asserted the right for the public to be protected from the '*risk* of eating what might be injurious to him', and suggested that experimental studies supported this stance.[29] Berry agreed with the prosecution. When he delivered his verdict, he found that both carcasses were 'unfit for human consumption'. Berry was convinced by the evidence that tuberculosis was transmissible from animals to humans by the ingestion of diseased meat. He criticised the process whereby the diseased parts were removed and carcasses sold, arguing that:

> There may be no appearance visible to the naked eye of the action of the tubercular bacillus in a practical part of the animal, and yet it may not improbably be there. The presence of the agent of the disease must precede the visible results of its action.[30]

In making the claim, Berry supported the stance adopted by those campaigning for more stringent meat inspection. He believed that the loss of a small

[27] *Tuberculous Meat*, pp. 154, 62.
[28] 'Important Trial Regarding Tuberculosis Carcases at Glasgow', *JCPT* ii (1889), p. 186; *Tuberculous Meat*, pp. 239, 280, 254, 241.
[29] 'Diseased Meat Inquiry at Glasgow', *Sanitary Record*, 15 July 1889, p. 528.
[30] 'Sale of Diseased Meat in Glasgow', *Lancet* i (1889), p. 1314.

proportion of the meat supply by condemning tuberculous carcasses 'is much to be regretted' but 'in weighing its importance I am of the opinion that it is sufficient to overcome those considerations in the interests of the public health'.[31]

'Tuberculous tyranny'

The trial stimulated further discussion not only on the threat from bovine tuberculosis but also on the issues of inspection, responsibility and compensation. Many were dissatisfied. The *Journal of Comparative Pathology and Therapeutics* summed up a common view among the medical and veterinary profession when it expressed disappointment that Berry had not delivered a clear verdict on the powers of meat inspectors to seize tuberculous carcasses.[32] The meat trade was incredulous and feared that Berry's decision would cause considerable anguish. In asserting the professional competence of butchers, the *Meat Trades' Journal and Cattle Salesman's Gazette* complained that much of the evidence had come from those who 'know nothing of cattle or their diseases' and had blindly accepted 'the dictum of a German theorist'. It stressed the value of practical experience over experimental studies or veterinary knowledge and concluded that 'the public suffer little danger in eating' meat from tuberculous cows. For the meat trade the science of how bovine tuberculosis was transmitted from animals to man remained contested. Not all, it would appear, were willing to accept the apparently disinterested authority of science when it clashed with their practical expertise. The *Meat Trades' Journal and Cattle Salesman's Gazette* therefore demanded further investigation to 'demonstrate fully the soundness of the proposition hypothetically raised regarding the communicability of consumption from the lower animals to man'.[33]

The trial, by reaffirming the dangers of bovine tuberculosis, directed public attention to the apparent 'wanton disregard of proper precautions to conserve the public health' and encouraged sanitary officials to embark on 'a crusade against the sale of tuberculous meat'.[34] In Glasgow, five additional meat inspectors were appointed. It was over-optimistically anticipated that these measures would see the city cease to be a centre for the sale of diseased meat.[35] Previously inactive local authorities started to prosecute butchers selling tuberculous meat

[31] 'Tuberculous Meat', *BMJ* ii (1889), p. 990.

[32] 'The Tuberculosis Meat Cases at Glasgow', *JCPT* ii (1889), p. 138.

[33] *MTJCSG*, 8 June 1889, p. 14.

[34] 'Tuberculous Meat', *Veterinary Record*, 9 Nov. 1889, p. 252; 'The Inspection of Meat', *JCPT* ii (1889), p. 144.

[35] GCA: Committee of health, 16 Nov. 1891, E1/20/13; *Report of the Operations of the Sanitary Department in the City of Glasgow for 1890* (Glasgow, 1891), p. 6.

and greater attention was directed at condemning carcasses that showed localised signs of the tubercles. As one contemporary commented, 'those gentlemen in this country who now condemn totally the flesh of slaughtered animals which exhibit localised tuberculosis' had 'only changed their practices' since the Glasgow trial.[36] This was clear in Liverpool where for the 'first time in memory . . . several animals had been wholly condemned, because they were suffering from tuberculosis'.[37] Across the country, there was a flood of cases as MOsH, encouraged by Berry's verdict, made direct links between the prevalence of tuberculosis in their towns and the sale of tuberculous meat. Whether the temporary drop in beef prices for 1889/90 indicated public anxiety is harder to determine, but by 1895 'seizures of meat are made every day by inspectors'.[38]

Butchers went on the defensive. Although the meat trade had been alarmed about the conditions reported in Glasgow, and had condemned the sale of tuberculous meat in the past, it felt threatened. As far as butchers were concerned it was not a question of whether bovine and human tuberculosis were related – many had largely accepted this by the late 1880s – but of the basis on which inspectors should condemn meat. They defended a localised model and were forthright in their views that 'the public health is not endangered by the eating of any animal only slightly affected'. Butchers complained that because it was 'impossible to diagnose tuberculosis in living animals' the disease was 'only found after their slaughter'.[39] Hence those who had purchased animals in good faith, believing them to be sound, felt aggrieved when they were prosecuted and the farmer escaped. Butchers' associations therefore pressed for measures to protect trade interests and called for compensation, revisiting issues that had been discussed in connection with rinderpest and pleuro-pneumonia in the 1860s and 1870s.[40]

The Glasgow trial also highlighted how divided opinion remained over tuberculous meat. The *BMJ* criticised the veterinary profession as not being the 'best qualified to judge as to the fitness of carcases for human food', whilst the *Veterinary Journal* responded with claims that doctors were 'far less trustworthy'.[41] This tension was revealed in Wigan. Inspired by the Glasgow trial, the MOH moved to prosecute a butcher for selling tuberculous meat. He argued

36 'Shall we Eat Tuberculous Meat?', *BMJ* i (1890), p. 865.

37 *MTJCSG*, 29 July 1889, p. 8.

38 Edith H. Whetham, 'Livestock Prices in Britain, 1851–93', *AHR* xi (1963), p. 29; *Sanitary Record*, 15 Dec. 1890, pp. 317–18; *MTJCSG*, 31 Jan. 1895, p. 674.

39 'Sale of Diseased Meat in Glasgow', *ibid.*, 4 May 1889, p. 7; *Royal Commission Appointed to Inquire into the Administrative Proceedings for Controlling the Danger to Man through the use as Food of the Meat and Milk of Tuberculosis Animals*, part ii, minutes of evidence (London, 1898), pp. 20–1.

40 *MTJCSG*, 20 July 1889, p. 5.

41 'Meat Inspection', *Veterinary Journal* ii (1889), p. 175.

that the high level of tuberculosis in the town was due to the practices of unscrupulous butchers, but the defendant used evidence from Alfredo Kanthack, bacteriologist at University College, Liverpool, to argue the disease was localised and that 'it has not yet been proved that the flesh of tuberculous animals is tuberculous'. For the *BMJ* such views were indefensible. It argued that 'no one can say absolutely, however glibly they speak of "localisation", that the germs of the disease have in any given case not commenced to be circulated in every part of the body'.[42] Here was the nub of the problem.

As the degree of danger presented by tuberculous meat was debated in the press, pressure was applied on the government to resolve the confusion. The question of the 'appreciable risk of [tubercular] disease' from meat was, for *The Times*, a 'momentous one'.[43] Doctors wholeheartedly agreed that investigation was essential into a disease that caused 'such terrible devastation on our race'. Many were worried that 'the scientific knowledge of the country in regard to the bacillus' was not sufficient to settle the question, feeding into growing fears about the scientific status of the country.[44] The meat press argued in a similar vein and lobbied parliament for protection. Butchers were alarmed that since the Glasgow trial local authorities had paid special attention to tuberculosis and consequently many had sustained considerable losses. They argued that no adequate proof existed that meat from tuberculous livestock was dangerous and called for the danger from tuberculous meat to be 'satisfactorily demonstrated'. For butchers this could only be achieved by a further government investigation to 'test the matter in such a scientific, tangible and reliable manner as to be above all cavil'.[45] All agreed that some form of investigation was vital; that the Glasgow trial had raised more questions than it had answered.

At the Local Government Board, George Buchanan as principal medical officer was keen to investigate. He had a long-standing interest in animal health and food hygiene, and under his influence the LGB had resolved to conduct epidemiological studies to determine 'the danger of infection from ingested tuberculous meat'. However, the Board of Agriculture tried to duck responsibility, noting that while it would offer any assistance it could, the danger had to be 'authoritatively settled' before it could act.[46] The financial implications of tackling bovine tuberculosis worried the Board of Agriculture, which was already anxious that the funds available to compensate stockowners

[42] 'The Meat Trade and Tuberculosis', *BMJ* i (1890), p. 1079.
[43] *The Times*, 26 Apr. 1890, p. 11.
[44] *BMJ* i (1889), p. 815; GCA: Joint subcommittee on the inspection of dead meat, 24 Sept. 1889, MP29.168.
[45] 'The Meat Trade and Tuberculosis', *BMJ* i (1890), p. 1079; 'Tuberculosis', *MTJCSG*, 7 Sept. 1889, p. 5.
[46] 'Tuberculosis in Cattle', *JCPT* iii (1890), p. 96.

for those diseases covered by the Contagious Diseases (Animals) Act were proving insufficient. As *The Lancet* realised, bovine tuberculosis represented a complicated equation between 'the respective values attaching to the lives of men and of beasts in the eyes of those who held the purse strings'.[47] Further investigation removed the immediate need for action at a time when the routine of office and overstretched government departments encouraged inertia and caution. Political, scientific and commercial differences merged in the appointment of a royal commission.

The dangers of meat?

The Departmental Committee on pleuro-pneumonia and tuberculosis had explicitly pointed to the dangers of bovine tuberculosis and the role played by meat. However, in the wake of the Glasgow trial the findings of the committee were no longer sufficient. As Hamlin has demonstrated, political decision-making in Britain as to the acceptance of germ theory and its implications for the water supply required an insistence on certainty, or at least a consensus.[48] The same might be said for bovine tuberculosis. In response to the confusion generated by the Glasgow trial, and under pressure from all sides, the government appointed a royal commission to provide this certainty. It focused on food safety because to deal with the disease in cattle would encounter staunch resistance from the strong agricultural lobby and cost too much. In an attempt to settle the difficult question of the extent to which diseased meat was dangerous, the commissioners investigated the arguments for total and partial seizure.

At first, sixteen expert witnesses, mostly drawn from the medical profession, were consulted. However, such was the contradictory nature of the evidence that it quickly became apparent that the findings were far from uniform and that the certainty being sought was elusive. As Hamlin has argued, commissions represented an ideal environment for expert disagreements.[49] Witnesses disagreed about whether tuberculosis was infectious, on the vectors of transmission, and the dangers of eating infected meat. With all the witnesses claiming that their findings were based on experimental studies and were hence objective statements, by the end of the first year the commissioners were no closer to solving the questions they had been set. Professional and expert opinion was so divided that it could not be relied upon to shape policy.

[47] *Lancet* i (1890), p. 484.

[48] Christopher Hamlin, 'Politics and Germ Theories in Victorian Britain: The Metropolitan Water Commissions of 1867–9 and 1892–3', in *Government and Expertise: Specialists, Administrators and Professionals, 1860–1919*, ed. Roy MacLeod (Cambridge, 1988), pp. 110–27.

[49] *Ibid.*, p. 111.

To overcome this it was decided to organise a programme of experimental research.[50]

Successive governments from the 1880s onwards had become increasingly willing to seek advice from members of the medical and scientific elite on issues of scientific research. If the state was a 'reluctant patron', it was nonetheless fully prepared to consult on, and fund, research into areas it considered of national importance, especially when these were related to health and agriculture.[51] Because the slow development of the symptoms of bovine tuberculosis precluded a style of epidemiological study favoured by the LGB, a laboratory approach was sought.[52] Despite the laboratory having an uncertain position in the construction of medical and veterinary knowledge in the late Victorian period, for the commissioners it embodied the authority they were looking for. At a time when it was being suggested that preventive medicine was becoming 'more and more lost in Bacteriology', the laboratory offered the public health movement a political authority based on science.[53] The commissioners therefore turned to the limited number of researchers who had experience of bacteriology and an interest in comparative pathology: John MacFadyean, Sidney Martin and German Sims Woodhead were appointed to conduct the commission's research. All had considerable knowledge of comparative pathology and laboratory work, with MacFadyean in the vanguard of re-establishing experimental work in veterinary pathology. In addition, all had close links with the commissioners and the LGB.[54]

The investigators (and their assistants) turned to tried and tested methods, using a mixture of feeding and inoculation experiments whereby tubercle infected material was fed or injected into a series of test animals. These methods fitted with an established tradition of research to test the communicable nature of disease. They involved an experimental framework familiar to British pathologists that was based on a style of bacteriology rooted in morbid histology and microscopic analysis. Inoculation experiments had been used in the late eighteenth century, and since the 1840s had become a staple in the pathological study of disease. They were considered more reliable in determining the presence of the tubercle bacillus than microscopic examination.[55]

[50] See Keir Waddington, 'The Science of Cows: Tuberculosis, Research and the State in the United Kingdom, 1890–1914', *History of Science* xxxix (2001), pp. 355–81.
[51] See Peter Alter, *The Reluctant Patron: Science and the State in Great Britain, 1850–1920*, transl. Angela Davies (Oxford, 1987).
[52] See Anne Hardy, 'On the Cusp: Epidemiology and Bacteriology at the Local Government Board, 1890–1905', *Medical History* xlii (1998), pp. 328–46.
[53] *JSM* iv (1896), p. 245; Christopher Hamlin, 'State Medicine in Great Britain', in *The History of Public Health and the Modern State*, ed. Dorothy Porter (Amsterdam, 1994), p. 150.
[54] Waddington, 'Science of Cows', pp. 355–81.
[55] See Edward Klein, 'Infectious Pneumo-Enteritis in the Pig', *Annual Report of the Medical Officer of the Local Government Board for 1877* (London, 1878), pp. 169–2802.

It was also common practice to use animals (principally guinea pigs) as living factories for bacteria, especially as even experienced experimenters encountered problems with artificially culturing bacteria. These methods formed the basis of many bacteriological studies and were used extensively in work on bovine tuberculosis. Little of the research for the commission was therefore new. Martin, Woodhead and MacFadyean duplicated much of the work Koch had pioneered in the 1880s, and repeated experiments similar to those undertaken at the Brown Animal Sanatory Institute for the LGB and for the veterinary department of the Privy Council.[56] Throughout, Martin, Woodhead and MacFadyean never questioned the zoonotic nature of tuberculosis, nor the bacteriological model of contagion. The state may have made an investment in experimental studies to settle how dangerous meat from tuberculous livestock was, but these were firmly rooted in accepted practices. They were not designed to be controversial or groundbreaking, and revealed the continuing limitations of bacteriology.

Although the announcement of tuberculin as a possible cure for tuberculosis temporarily directed attention from the problem of bovine tuberculosis, once tuberculin was shown to have a limited curative value, renewed interest was expressed in combating bovine tuberculosis as a means of removing one of the sources of contagion. Whilst the researchers collected evidence, speculation continued as to the extent of the danger represented by meat. Sanitary inspectors were advised to 'err on the safe side' until 'the question is more definitely settled'.[57] This did not stop advocacy for more systematic meat inspection as criticisms of existing arrangements mounted. Protests were voiced against the tendency for meat inspection to be placed in the hands of 'persons having little or no knowledge of animal diseases, and hence unfitted to discharge their function'.[58] It was also argued that sanitary officials required more stringent powers to deal with tuberculous meat. It was anticipated that the commission would lead to legislation for better inspection and the eradication of the disease. However, the commission also offered an opportunity for inaction. In Birkenhead, for example, a magistrate refused to condemn an 'obviously' tuberculous carcass 'in view of the fact that a Royal Commission was at present considering this very point'.[59] The Board of Agriculture also used the commission as an excuse to delay when it came to measures to slaughter tuberculous cattle.[60]

[56] *Departmental Committee . . . into Pleuro-pneumonia and Tuberculosis*, minutes of evidence, pp. 254–5; *Annual Report of the Agricultural Department of the Privy Council Office, on the Contagious Diseases, Inspection, and Transit of Animals for the Year 1888* (London, 1889), pp. 37–45.

[57] George Reid, *Practical Sanitation: A Handbook for Sanitary Inspectors and Others Interested in Sanitation* (London, 1895), p. 254.

[58] Henry Behrend, *Cattle Tuberculosis and Tuberculous Meat* (London, 1893), p. 83.

[59] 'Tuberculous Beef', *Sanitary Record*, 16 Mar. 1891, p. 450.

[60] 'Tuberculosis in London', *Veterinary Record*, 25 Feb. 1893, p. 470.

The commissioners finally delivered their report in 1895. They reasserted a contagious model of tuberculosis and outlined what quickly became the British model. Hereditary ideas were dismissed, although notions of susceptibility were maintained, and the veterinary understanding of the disease was confirmed. Martin classified cases of tuberculosis according to the extent of the lesions, referring to 'mild', 'moderate' and 'generalised' cases.[61] The latter was reserved for the most serious cases, which Martin believed represented a small proportion of the total number of animals killed. His taxonomy was new, but his understanding of the disease was influenced by studies for the veterinary department of the Privy Council.[62] Martin dismissed what he saw as 'egregious irregularities' in his experimental results when not obviously diseased flesh from infected cattle gave rise to tuberculosis in test subjects. Variable results were frequent in bacteriological studies, but Martin was willing to overlook these irregularities because he was convinced that tuberculosis was not a disease in which the infective agent 'is present in all parts of the animal'. He therefore concluded that infection was not caused by 'any infective material present in the muscular tissue itself'. Instead, Martin blamed unhygienic or inadequate butchering techniques, transferring the blame to an already beleaguered meat industry. In adopting this approach he played down meat's role in infection and the need for 'total seizure' advocated by many MOsH as he believed that muscular tissue rarely showed signs of lesions, a view confirmed by other researchers in Europe. Only in advanced cases, when the cow was obviously tubercular, did Martin consider meat dangerous. MacFadyean and Woodhead agreed.[63]

Martin's assessment was accepted by the commissioners as the most convenient in part because it appeared to justify the status quo. As a result, the commissioners said little that was new. They dismissed findings that did not fit generalities and asserted that with 'sufficient discrimination and care . . . a great deal of meat from [tuberculous cattle] . . . might without danger be consumed'. Martin's 'egregious irregularities', which justified a policy of total seizure, were conveniently ignored because to accept a generalised model of infection would have serious implications for the meat trade and food supply. By supporting the notion of local infection, a policy of 'partial seizure' for all but advanced cases could be justified; a move that was less damaging to farming and the meat industry and less expensive.[64] After four years of experiments, the commissioners' findings endorsed existing concerns about

[61] *Royal Commission Appointed to Inquire into the Effect of Food Derived from Tuberculous Animals on Human Health*, part ii, inquiry ii (London, 1895), pp. 10–31.

[62] See *Annual Report of the Agricultural Department of the Privy Council Office, on the Contagious Diseases, Inspection, and Transit of Animals for the Year 1888* (London, 1889), pp. 37–45.

[63] *Royal Commission . . . into the Effect of Food Derived from Tuberculous Animals*, inquiry ii, pp. 10–31.

[64] *Ibid.*, part i, pp. 10–17.

the pathogenicity of bovine tuberculosis and the potential for meat to transmit the disease. However, at the same time, by adopting a localised model, they went against a growing body of sanitary officials who thought that any sign of the disease should result in the entire carcass being condemned.

Compensation and control

The publication of the commission's report marked the zenith of interest in tuberculous meat. Although doctors and veterinarians continued to point to the alarming frequency with which tuberculosis was transmitted through meat, the commission's findings that milk represented a greater danger attracted increasing attention (see chapter 9). However, the commissioners' report left many questions unresolved. Part of the problem was that the commission, in focusing on the science of bovine tuberculosis and infection, had not put forward any practical recommendations, pointing to the often uneasy relationship between science and prevention policies. As such, the commissioners did not produce the answers doctors, veterinarians or butchers were looking for and did little to end the existing chaotic system of meat inspection. The commission was therefore attacked as 'an admirable machinery for the production of delay' that left few satisfied.[65]

The meat trade saw the commissioners' report as 'practically useless' because it had left the exact danger from tuberculous meat vague and had failed to address the thorny issue of compensation.[66] Compensation had become a major 'bone of contention' between butchers and sanitary authorities in condemning diseased meat. The principle of compensation for diseased livestock had been established through successive Contagious Diseases (Animals) Acts as a mechanism to encourage the eradication of epizootics. Butchers wanted to see the principle extended to reduce what they believed was the unfair burden placed on the meat trade from sanitary officials condemning tuberculous meat. Although livestock farmers saw animal disease as an occupational hazard, butchers wanted to minimise the risk of buying tuberculous cattle, which were difficult to detect before slaughter. In a petition to the LGB and Board of Agriculture, the National Federation of Butchers and Meat Traders' Associations pressed the point that the seizure of diseased meat without compensation put butchers under considerable financial strain 'for what they cannot possibly avoid' at a time when the industry was beginning to feel the effects of competition from imported meat. The Federation returned to familiar arguments about the damaging effect condemnation had on business and pressed for compensation. However, not all were convinced by these arguments. Butchers

[65] 'Royal Commission on Tuberculosis', *The Times*, 27 Apr. 1895, p. 9.
[66] *Royal Commission . . . into . . . Controlling the Danger to Man through the use as Food of the Meat and Milk of Tuberculosis Animals*, minutes of evidence, p. 18.

were attacked for exaggerating the extent to which meat was seized, whilst others shared the view of livestock farmers that the seizure of diseased meat was a legitimate 'business risk' that did not require compensation or intervention in the workings of the market.[67] Local authorities were also uneasy: many had watered down measures to award compensation under the various Contagious Diseases (Animals) Acts, anxious not to burden the rates.[68] The whole issue of compensation was fraught with problems, not least who should pay.

In an attempt to answer some of the questions the first commission had left unresolved, a second commission was appointed in 1896. It was anticipated that this new commission would draft the 'thoroughly comprehensive and workable scheme' that many believed vital to combat bovine tuberculosis.[69] From the outset it was clear that the first commission had not created a consensus on the dangers of tuberculous meat. Although the new commission endeavoured to focus on practical measures to limit bovine tuberculosis, differences as to the extent to which tuberculous meat was considered a danger to human health continued to emerge. As in other areas, doubts were expressed about the value of science in determining the extent of the danger. Those giving evidence for the agricultural industry were sceptical of 'scientific evidence', asserting their credentials 'as practical men', pointing to the ongoing tensions between notions of expertise and scientific knowledge.[70] Although the commissioners remained convinced that the findings of the first commission were valid, they did side with the view that the threat from tuberculous meat had been exaggerated. This stance shaped the commissioners' conclusions and saw greater emphasis placed on the dangers of milk.

One of the key issues addressed was the nature of meat inspection. Witnesses pointed to the ongoing chaos in inspection. Representatives of the farming and butchers' lobby complained about unfairness and a lack of uniformity, often implying that MOsH were at fault, unnecessarily condemning carcasses. The National Federation of Butchers and Meat Traders' Associations explained that the meat trade did not desire 'to shirk inspection',

> but that there should be co-operation with municipal authorities and with practical men for the protection of the community, and also for the protection of honest meat salesmen and retailers and butchers, because we hold they ought to be protected from prosecution.[71]

[67] 'The Seizure of Diseased Meat and Butchers' Claims to Compensation', *BMJ* ii (1896), p. 319; GCA: Report on the prevention of tuberculosis, 23 Dec. 1895, LP1/18.
[68] See *Annual Report of the Veterinary Department of the Privy Council Office for the Year 1878* (London, 1879), p. 7.
[69] *Veterinary Journal* xlii (1896), p. 253.
[70] See *Royal Commission . . . into . . . Controlling the Danger to Man through the use as Food of the Meat and Milk of Tuberculosis Animals*, minutes of evidence, p. 204.
[71] *Ibid.*, part ii, p. 20.

Many called for clear guidelines, something the commissioners were keen to encourage. Yet, despite a general sense that meat inspection needed to be improved, divisions appeared as to how this was to be achieved, reflecting the different interpretations of how dangerous meat from tuberculous animals was.

However, the most common grievance expressed was the risk farmers and butchers were felt to shoulder. Witnesses agreed that some form of compensation was necessary in the 'public interest', as the existing situation was felt to create 'a premium for concealment'. Farmers and butchers, anxious to minimise the risk of purchasing tuberculous cows, drew attention to how the burden fell on them: in Lancashire alone it was estimated that with thirty-five to forty beasts being condemned per week farmers were losing £25,000 per annum. Comparisons were drawn with the compensation provisions under the Contagious Diseases (Animals) Acts. Notwithstanding problems with enforcing the Acts, especially when it came to payment, it was felt that compensation did much to 'reveal' the true extent of disease by encouraging farmers and butchers to come forward. It was seen as 'a kind of bribe to the owner of the animal to give notice of the disease', so that it could be slaughtered and the meat destroyed.[72] However, few suggested that compensation should be universal. Most giving evidence agreed that it should only cover cases where it had been impossible to detect the disease before slaughter, or at least where the sanitary authority had not been able to determine the presence of any tubercles in dead meat. All could agree that compensation should be funded from the national Exchequer, a need underlined by the defects in the system of local administration under the Contagious Diseases (Animals) Acts.[73]

The second commission put forward a package of reforms designed to limit infection from bovine tuberculosis that it felt mirrored the views of public health officials and veterinarians on how to combat the disease. In a move to reassure the public, the commissioners played down the risk from infected meat, reflecting a growing feeling that aerial infection was the main source of contagion.[74] If the danger from diseased meat was felt to be exaggerated, milk was now shown to represent the greatest danger from food to health. The identification of milk merged with wider concerns about its role in the spread of disease and growing anxiety about the health of the nation that were to coalesce around fears of physical deterioration. However, although the commissioners considered that the risk from meat had been overestimated, they still argued that there was a degree of danger. Worried about discrepancies in meat inspection practices, the commissioners sought to end the existing chaotic system. They therefore outlined guidelines for uniform standards of inspection to protect the consumer who could not be expected to identify

[72] *Ibid.*, pp. 86, 227, 213.
[73] *Ibid.*, pp. 57, 73, 103, 224.
[74] *Ibid.*, part i, pp. 1–24.

diseased meat, and called for the appointment of qualified meat inspectors to replace existing officers, many of whom were believed to lack any real skill in identifying diseased meat. Here the commissioners were influenced by European practices. It was hoped that improved meat inspection would also reduce the risk from other food-borne diseases.[75]

The commissioners were less clear about the issue of compensation. They accepted butchers' protests that seizure was a burden, but concluded that the risk had been overestimated. The commissioners were worried that compensation would open the floodgates for claims in other areas, whilst concern was voiced about possible deception. The commissioners therefore rejected compensation in favour of a system of mutual trade insurance schemes that transferred financial responsibility back to the butcher and mirrored earlier suggestions that the consequences of non-epizootic cattle diseases should be met by the trade.[76] At a time of frequent recessions, and given repeated attempts at retrenchment in domestic expenditure, the problematic issue of compensation was too expensive to negotiate successfully.

The measures outlined by the commissioners were designed to be palliative, as with much public health work, and to operate at a local level. Although they recognised the importance of eliminating the disease in cattle, the commissioners did not recommend measures intended to secure the widespread eradication of bovine tuberculosis. The need for farmers and cattle dealers to act was therefore sidestepped, a move that parallels food producers' successes at avoiding regulation in other areas of food legislation.[77]

Conclusions

The report of the second commission did much to allay the alarm created by the first. However, neither commission provided the certainty that many were looking for despite their reliance on experimental science. Although the state had employed the emerging authority of pathological and bacteriological studies to define a British model that bovine tuberculosis was a danger to health, not all were convinced, pointing to the different values assigned to the science commissioned. Even writers on public health continued to feel that 'the question of allowing as food supply the flesh of animals affected with Tuberculosis . . . is a vexed one'.[78] The *Veterinary Record* remained critical, blaming the commissions for the ongoing 'confusion and ignorance' that

[75] *Ibid.*, pp. 6–10, 21.
[76] *Ibid.*, pp. 11–12.
[77] See Jim Phillips and Michael French, 'Adulteration and Food Law, 1899–1939', *Twentieth Century British History* ix (1998), p. 354.
[78] O. Andrews, *Handbook of Public Health Laboratory Work and Food Inspection* (London, 1901), p. 24.

surrounded the issue.[79] There were further practical problems. Compensation and who should pay were just as contentious after the commissions as they had been in earlier schemes to combat epizootics. A chorus of disapproval from farmers and butchers followed both commissions. They remained sceptical of the science employed and were dissatisfied that neither commission had endorsed compensation. In the agitation that followed farmers and stockowners became increasingly intransigent. Fears were expressed about the hegemony of science in defining diseased meat which appeared to threaten the competency of butchers. Continued pressure was applied by the National Federation of Butchers and Meat Traders' Associations on the LGB and the Board of Agriculture to secure better treatment for butchers under the guise of protecting the public. Local associations of butchers, frustrated that compensation had not been endorsed, took matters into their own hands and started to insist on guarantees from traders and stockowners. By 1908, the question of a warranty appeared to threaten a 'trade conflict' as farmers, dealers and stockowners resisted pressure from the National Federation for a national system of guarantees.[80]

Public and professional concerns were also stimulated by the second commission. Fears about tuberculosis had been rising. The formation of the Association for the Prevention of Consumption in 1898 marked the start of a national campaign to eradicate the disease in Britain that was to shift attention to pulmonary tuberculosis.[81] Part of an international movement and a growing sense that Britain was 'very much behind in these matters', the Association was convinced that the disease was communicated from person to person and from animals to humans. It was hence concerned with all forms of tuberculosis.[82] However, alarmed that levels of infection from bovine tuberculosis had not fallen, unlike other forms of the disease, the Association worked to raise public awareness about the sources of infection from cows and to promote the eradication of the disease in cattle as part of its general campaign against consumption.[83] Following the commission, the SMOH also started to discuss 'what further administrative or other measures can be taken to check the spread of tuberculosis'.[84] Local conferences were held to determine what practical measures could be implemented as the drive to prevent the sale of meat and milk from tuberculous animals as part of the crusade against consumption became common currency.

[79] *Veterinary Record*, 10 Sept. 1898, p. 141.
[80] *Ibid.*, 25 Feb. 1899, p. 503; 'Butchers' Warranty', *ibid.*, 24 Oct. 1908, p. 259.
[81] Linda Bryder, *Below the Magic Mountain: A Social History of Tuberculosis in Twentieth-Century Britain* (Oxford, 1988), p. 15.
[82] *National Association for the Prevention of Tuberculosis Bulletin* xi (1948).
[83] *Veterinary Record*, 24 Dec. 1898, p. 349.
[84] Wellcome: SMOH council minutes, 11 Nov. 1898, SA/SMO/B.1/8.

Although the two commissions had identified milk as the main source of infection, and hence the greater danger to the public, alarm about tuberculous meat was not defused. That milk was an important agent of contagion was not contentious; the view that the role played by meat had been exaggerated was harder to accept, especially at a local level as it had the potential to damage attempts by public health officials to clean up the meat trade. These efforts to improve inspection and establish public abattoirs were directly influenced by the findings of the royal commissions, prompting renewed efforts to extend meat inspection, raise standards and regulate the meat trade.

7

Experimentation and the British model

By the start of the twentieth century, the importance of animals to the transmission of disease to humans had gained considerable ground. Issues relating to animal health and human disease were discussed by professional bodies, which started to cover veterinary issues at their annual meetings. At the same time, the 1890s had seen growing concern about food safety as the number of local food scares proliferated. Much of this anxiety focused on tuberculous meat and milk. The two royal commissions had helped fashion a British model that bovine tuberculosis was a threat to public health; that meat and milk were important vectors of transmission. Although the exact nature of the danger remained uncertain, most agreed that a better system of meat and milk inspection was crucial to protect the public. Problems remained, particularly over the questions of compensation and uniform inspection, but few voiced doubts that bovine tuberculosis had to be tackled.

The sense that bovine tuberculosis had to be checked built on a view that experimental pathology and clinical observation had removed all doubt that the disease could be transmitted to man via meat and milk. However, this assessment was rocked by Robert Koch's speech at the 1901 British congress on tuberculosis. Koch's claims that the disease did not represent any real danger shook the public health machinery. He challenged assumptions about the scientific classification of the disease, the role of bovine tuberculosis in public health, and its role as an area where progress had been made in the control of tuberculosis. Once more the British state responded with the appointment of a royal commission, the third since 1891, as debate focused on the morphological and biological characteristics of the tubercle bacillus and what measures, if any, were needed to protect the public from infected meat and milk. Renewed faith was placed in the value of experimental studies, this time to prove Koch wrong.

The third commission was part of an anxious flurry of international investigations stimulated by Koch's claims. It put together a combination of scientific, social and political forces to defend the British model that human and bovine tuberculosis were the same disease, 'and that no difficulty is experienced in transmitting the malady'.[1] In the process the science of bovine tuberculosis was defined, experimental processes refined, and a research infrastructure

[1] Thomas Brown, *The Complete Modern Farrier: A Compendium of Veterinary Science and Practice* (Edinburgh, 1900), p. 527.

established. At no point was the idea that bovine tuberculosis was a serious public health issue questioned. If the earlier royal commissions were mechanisms for delay, from the start the third commission sought to defend a campaign against consumption that saw bovine tuberculosis as one area in which tuberculosis could be fought successfully.

The 1901 London congress

By the end of the nineteenth century few pronouncements on tuberculosis did not include some reference to the bovine form of the disease. Better meat inspection was considered vital to prevent the sale of diseased meat; more rigorous controls of the dairy industry necessary to ensure a pure milk supply. How this was to be achieved, however, and how bovine tuberculosis should be eradicated from herds, remained uncertain despite the findings of the earlier royal commissions. The 1901 British congress on tuberculosis in London was therefore designed to exchange information on methods to combat tuberculosis at an international level. It was part of a series of meetings on the disease encouraged by the report of the second royal commission on tuberculosis and the resulting crusade against consumption. Organised by the British National Association for the Prevention of Tuberculosis (which had expanded rapidly following the commission) and other interested parties, it was considered the 'most important step that has yet been taken in this country with a view to applying in a practical manner the information and experience already acquired as to methods for stamping out consumption'.[2] The congress was to be a venue for public education, as well as for the exchange of scientific information: it was widely advertised and prominent public figures were invited. The main thrust was to be on pulmonary tuberculosis, and, as Rosenkrantz has suggested, there was no reason why bovine tuberculosis should become an 'inflammatory issue'.[3] More was known about bovine tuberculosis than about other forms of the disease, and most European states had developed measures to combat the sale of tuberculous meat and milk. However, the congress was to renew and reinvigorate debate about the disease.

Of all the speakers, it was Koch's paper that generated most discussion. Koch was widely recognised as a brilliant investigator. His identification of the bacillus had helped transform 'the problem of prevention and treatment' from something nebulous to a concrete proposition and made him the pre-eminent authority on tuberculosis.[4] His earlier work and international status ensured

[2] *BMJ* i (1900), p. 1110; GRO: LGB circular, 1900, BC/C/6/22.

[3] Barbara G. Rosenkrantz, 'Koch's Bacillus: Was there a Technological Fix?', in *The Prism of Science*, ed. Edna Ullmann-Margalit (Boston, 1986), p. 156.

[4] Cited in Linda Bryder, *Below the Magic Mountain: A Social History of Tuberculosis in Twentieth Century Britain* (Oxford, 1988), p. 17.

that his pronouncements attracted considerable attention. His paper at the conference therefore 'fire[d] off a bombshell'.[5]

In his address, Koch rejected established notions of the aetiology of bovine and human tuberculosis by proclaiming that 'human tuberculosis differs from bovine'.[6] He based his claims on second-hand clinical and epidemiological evidence rather than on the bacteriological studies that had made his name. Using this evidence, Koch argued that bovine and human tuberculosis were not one disease, but two. Others had already pointed to differences in the size of the bacilli and had raised doubts about the morphological similarities.[7] That the tubercles in cattle tended to be concentrated in the pleura rather than the lung (as in man) added weight to suggestions that differences existed between the two bacilli. However, Koch went further. Drawing on a morphological tradition in German pathology, he built on the findings of the American bacteriologist Theobald Smith who had called attention to the differences between the two bacilli.[8] For Koch this was not a sudden departure. He claimed that he had always expressed 'reserve' with regard to the identity of bovine and human tuberculosis. This was not entirely accurate: in his first substantial publication on the aetiology of tuberculosis he had claimed that the diseases were identical. His earlier work had also been based on a unified conception of tuberculosis.[9]

More importantly, Koch threw doubt on 'whether man is susceptible to bovine tuberculosis' by arguing that 'the infection of human beings is but a very rare occurrence'. Koch had already speculated in 1884 that bovine tuberculosis was less virulent for humans and had become convinced after examining post-mortem evidence that suggested primary intestinal tuberculosis was rare. This was supported by selective use of evidence from experiments at the Berlin veterinary school conducted in Koch's absence whereby nineteen young cows were injected with human bacilli.[10] Koch concluded that because these cows had not been susceptible to human bacilli, the converse must also be true. He backed up this view by noting that meat and milk from tuberculous cows had been used for years without any obvious epidemiological evidence that they caused infection.[11]

[5] 'The Congress on Tuberculosis', *Nature* lxiv (1901), p. 327; *Lancet* i (1904), p. 492.

[6] Robert Koch, 'An Address on the Fight against Tuberculosis', *BMJ* ii (1901), pp. 190–1.

[7] 'On the Relation of the Tubercle Bacilli to Artificial Tuberculosis', *Annual Report of the Medical Officer of the Local Government Board for 1883* (London, 1884), pp. 177–85; *Annual Report of the Agricultural Department of the Privy Council Office, on the Contagious Diseases, Inspection, and Transit of Animals for the Year 1888* (London, 1889), p. 45; *Veterinary Record*, 29 Nov. 1902, p. 351.

[8] *Transactions of the Association of American Physicians* xi (1896), pp. 75–93; *Journal of Experimental Medicine* iii (1898), pp. 451–511.

[9] Koch, 'Fight against Tuberculosis', pp. 190–1.

[10] 'Human and Bovine Tuberculosis', *BMJ* ii (1904), p. 355.

[11] Koch, 'Fight against Tuberculosis', pp. 190–1.

Although opinions had varied on the frequency with which bovine tuberculosis was communicated to humans, Koch's paper flew in the face of conventional thinking. He went on to claim that 'I should estimate the extent of the infection by the milk and flesh of tuberculous cattle, as hardly greater than that of hereditary transmission, and I therefore do not deem it advisable to take any measures against it.'[12] Experimental studies in France, Germany and Denmark in the 1890s had begun to question the virulence of tuberculous meat, a point accepted by the second royal commission. However, if they downplayed the part meat played in infection, they had not dismissed it and continued to suggest that tuberculous meat should be condemned as unfit.[13] Koch's claims went much further. For Koch, the effort and money directed at bovine tuberculosis would be better directed at a single-minded campaign against pulmonary tuberculosis, which he considered the main source of tuberculous infection.[14] In the furore that followed, Koch's admission that the risk from bovine tuberculosis was 'not absolutely non-existent' tended to be overlooked.[15] Koch had defined bovine tuberculosis in 1882; now he issued a challenge to the science and public health policies constructed around the belief that bovine tuberculosis was dangerous.

Why Koch made these claims is uncertain. At the time, some speculated that he had been pressurised by the German agrarian community alarmed about the adverse impact of bovine tuberculosis on agriculture.[16] By minimising the perceived risk from bovine tuberculosis, and by advocating a single-minded campaign against pulmonary tuberculosis, his views reflected growing international efforts to combat tuberculosis through individual responsibility, education and open-air treatment. Given Germany's much admired system of meat inspection Koch may also have felt that a focus on the bovine form of the disease was unnecessary. Koch might also have been trying to recover his status. The tuberculin fiasco had undermined his reputation, whilst his work on rinderpest and other cattle diseases in Africa in the 1890s had produced very meagre results.[17] Koch was not only issuing a challenge to a unified conception of tuberculosis he was doing so to give himself a fresh start that would bring his research programme back to the centre of international work on tuberculosis. Certainly, the 1901 conference once more pushed Koch to the forefront of public and scientific attention.

At the congress, John MacFadyean, head of the RVC and former investigator for the first royal commission, reluctantly responded with a

12 *Ibid.*

13 See Robert Ostertag, *Handbook of Meat Inspection* (London, 1904), pp. 636–7.

14 *Transactions of the British Congress on Tuberculosis* i (London, 1902), pp. 34–5.

15 Koch, 'Fight against Tuberculosis', pp. 190–1.

16 *Veterinary Record*, 3 Aug. 1901, p. 85.

17 Lise Wilkinson, *Animals and Disease: An Introduction to the History of Comparative Medicine* (Cambridge, 1992), pp. 195–7. For Koch's work in Africa, see Paul F. Cranefield, *Science and Empire: East Coast Fever in Rhodesia and the Transvaal* (Cambridge, 1991).

condemnation of Koch's paper. Although he had made similar, if less strident claims two years earlier at a meeting of the National Veterinary Association, he was alarmed that he was forced to 'formally prove' that the disease was the same in animals and humans. MacFadyean questioned Koch's evidence and proceeded to refute each point, hoping that the weight of counter evidence would be sufficient.[18] MacFadyean was keen to defend existing ideas and responses to bovine tuberculosis, but he was not alone in his views. A chorus of alarm and confusion followed. Even those sympathetic to Koch were uneasy about the limited evidence he had put forward.[19] Vigorous protests were made by leading bacteriologists and attempts were made to persuade Koch to withdraw his claims. However, he remained adamant, convinced of his own infallibility.[20]

Attention transcended the scientific community. The statistical importance of tuberculosis in mortality, the fear and stigma of contracting the 'dread disease', and the increasing availability of scientific information in the public dominion ensured that Koch's paper attracted considerable popular coverage. With the conference widely publicised, the public were drawn into the controversy. Questions were raised about the science of bovine tuberculosis and the campaign against consumption at a time when science was being incorporated into discussions about Britain's economic and political position in the world. Because of claims that the common identity of bovine and human tuberculosis 'is a thing to be accepted as a fundamental principle', a point confirmed by two royal commissions, the responses to Koch's paper ranged from 'incredulity' to 'astonishment' to anger.[21]

The prospect of discussing bovine tuberculosis was dramatically altered by Koch's paper. Whether or not tuberculosis could be communicated from cows to humans suddenly became a 'vexed question' where it had previously been accepted.[22] If Koch was right existing measures to prevent the spread of bovine tuberculosis would have to be repealed. Whether he was right or not was considered of less importance than the threat his ideas posed to the campaign against tuberculosis. Although many had already accepted that meat and milk were not the main source of infection, and that the pulmonary form of the disease was the chief killer, Koch's views struck at the one area of preventing tuberculosis that was felt to promise a practical way of reducing levels of the disease. Acknowledging that a morphological difference existed between the bovine and human forms of the tubercle bacilli was one thing, and had been

[18] John MacFadyean, 'Tubercle Bacilli in Cows' Milk as a Possible Source of Tuberculous Disease in Man', *JCPT* xiv (1901), pp. 215–24.

[19] George Newman, 'Tuberculous Milk and Meat', *Food and Health*, 19 Sept. 1901, p. 22.

[20] Edgar Crookshank, *Bacteriology and Infectious Diseases* (London, 1896), pp. 389–91.

[21] *Transactions of the International Congress on Hygiene and Demography* 2 vols (London, 1891), vol. ii, p. 185.

[22] *Sanitary Record*, 8 Aug. 1901, p. 126.

widely accepted; arguing that this meant that the bovine disease was no threat to human health was another. If some were prepared to suspend judgement 'in the face of such an authority as Koch', no one 'seemed inclined to quietly accept the new doctrine'.[23] Many found it hard to believe that whereas 'the labours of hundreds of workers' since 1882 had 'produced nothing in serious conflict with the conclusion that human and bovine tuberculosis were identical diseases', and had only confirmed its common identity, Koch could argue differently.[24] Notwithstanding the virtual reverence in which Koch was held, the *Journal of Comparative Pathology and Therapeutics* counselled that 'there are no Popes in science' and cautioned that the 'mere *ipse dixit* of one man, however eminent he may be, is not sufficient to upset convictions arrived at after examination of the whole available evidence'.[25] However, the journal misjudged the situation. It was precisely because of Koch's standing that his views had such an impact, throwing the medical and veterinary communities into confusion, and highlighting clinical and observational discrepancies in studies of bovine tuberculosis.

The confusion generated by Koch had an immediate impact and did 'an enormous amount of damage' to attempts to prevent the sale of tuberculous meat and milk.[26] Butchers welcomed Koch's views and used them to campaign for a relaxation of restrictions on diseased meat. Some sanitary authorities began to question the need for strict controls on meat and milk, despite calls by the LGB that pending investigation 'there should be no relaxation on their part or on that of their offices in the taking of proper measures for dealing with milk from tuberculous cows and with tuberculous meat'.[27] The LGB was alarmed, having become increasingly interested in the problem of food safety. It noted that Koch had 'raised questions for the Board of no small administrative difficulty', especially as at a local level efforts had been made after the second royal commission to tightened meat and milk regulation.[28] Opinion increasingly came to favour the need for studies to determine the accuracy of Koch's claims.

Amid a flurry of international activity to determine the accuracy of Koch's statements, the LGB secured the appointment of a third royal commission to investigate whether tuberculosis in animals and humans was the same, whether animals and man could be 'reciprocally infected', and under what conditions transmission occurred. Questions about whether bovine tuberculosis was localised or generalised, which had until 1901 shaped debate, were pushed to

[23] 'The Tuberculosis Congress', *Veterinary Record*, 27 July 1901, p. 46.

[24] MacFadyean, 'Tubercle Bacilli', p. 216.

[25] *JCPT* xiv (1901), p. 364.

[26] *Report of the Second International Congress on Alimentary Hygiene* i (1910), p. 49.

[27] CLRO: Public health committee, 23 July 1901.

[28] *Annual Report of the Medical Officer of the Local Government Board* (London, 1903), p. xxix.

one side by Koch's announcement with a tacit agreement that it could be both. Koch's views had raised fundamental questions about bovine tuberculosis and once more faith was placed in the authority of the laboratory to determine the answers.

Defending the British model

The men appointed to head the commission were felt to 'possess the confidence of the medical profession and the public'.[29] German Sims Woodhead, MacFadyean and Sidney Martin were all appointed commissioners, having served as investigators for the first commission. Although they were keen to point out that their work was conducted with 'open minds', most of the commissioners 'were declared hostile critics of Dr. Koch before any experiments were undertaken'.[30] The most Martin was prepared to admit was that there were 'some differences' in the manner in which tuberculosis affected humans and cows but continued to feel that 'no absolute distinction exists'.[31] In seeking to refute Koch, the commissioners worked to confirm the British model of the essential identity of the tubercular disease in humans and in cows and its 'communicability'.[32] Too much, politically and scientifically, had been invested in proving Koch's initial statement that bovine and human tuberculosis were connected; too many researchers in Britain had placed their faith in the fact that bovine tuberculosis was dangerous to humans, for any other stance to be taken.

However, the approach adopted by the commission was not unprecedented. Koch's ideas had already been called into question in a number of areas. Claims in 1890 that tuberculin was a cure for tuberculosis had quickly proved an embarrassment for Koch, although pockets of support for tuberculin as a form of treatment continued in Britain. Closer parallels exist with the British reaction to Koch's 1885 statements on cholera following a study of the epidemic in Egypt two years earlier. In identifying a 'cholera germ', Koch gave support for quarantine measures to control the spread of the disease. His conclusions ran counter to British arguments that advocated a sanitary solution and dismissed quarantine as unnecessary. Just as with bovine tuberculosis eighteen years later, Koch's claims threatened British public health policy. With Britain having 'political and economic motives for resisting germ-theory explanations

[29] *Sanitary Inspectors' Journal* vii (1901–2), p. 105. For the membership of the commission see Keir Waddington, 'The Science of Cows: Tuberculosis, Research and the State in the United Kingdom, 1890–1914', *History of Science* xxxix (2001), p. 364.

[30] 'The Transmission of Consumption by Animals to Man', *Cowkeeper and Dairyman's Journal*, Mar. 1907, p. 193.

[31] 'Tuberculosis', in *A System of Medicine*, ed. Thomas Clifford Allbutt and Humphry Davy Rolleston (London, 1906), p. 262.

[32] Edward Willoughby, *Milk: Its Production and Uses* (London, 1903), p. 170.

of cholera', official scientific evidence was marshalled that served to counter Koch's ideas and defend Britain's stance on quarantine which was under attack in the international arena.[33] As was to happen in 1901, a team of researchers had been selected that was already unsympathetic to Koch's theories. The royal commission on tuberculosis of 1901 to 1911 was to follow a similar pattern. From the start, it was designed to prove Koch wrong and to defend British public health responses to bovine tuberculosis. However, whereas the earlier investigation into cholera had used epidemiology to attack assumptions based on germ theory, the royal commission used the same theoretical and experimental framework that Koch had employed, rejecting a commission of expert witnesses in favour of a programme of state-sponsored bacteriological and pathological research.

In a climate in which Germany was increasingly being cast as having a scientific structure that was to be admired and perhaps adapted to Britain to combat the perceived scientific shortfall of British teaching hospitals, the royal commission represented the large-scale state investment in scientific research that a growing number of doctors believed essential to 'modern' medicine. At another level, the commission can also be understood as an expression of mounting antagonism between Germany and the British Empire, and attempts to reassert Britain's scientific standing. However, this anxiety did not match the popular faith placed in the value of science. Science had acquired an intellectual and moral authority, and it was to science that the third commission turned for answers.

Just as with the first commission, work was constructed around a British approach that continued to fuse experimentation with observation. With research straddling public health and a methodology derived from bacteriology and veterinary pathology, the commission created a research infrastructure that was not linked to teaching hospitals or university laboratories, despite their work in identification and diagnosis. Many of the existing facilities for analysis were overwhelmed by routine work and lacked the financial and technical resources necessary to carry out the large-scale research envisaged by the commissioners.[34] Instead of relying exclusively on outside researchers working through their institutions (a policy adopted by the LGB), research staff were appointed and two laboratories were established at Blythwood and Walpole farms near Stansted in Essex with a further laboratory at Royalcot. These were comparable to the agricultural research stations established in the United States.[35] Through the work of the royal commission, research into bovine tuberculosis (and later human tuberculosis) was institutionalised.

[33] See Mariko Ogawa, 'Uneasy Bedfellows: Science and Politics in the Refutation of Koch's Bacterial Theory of Cholera', *BHM* lxxiv (2000), pp. 671–707, 686.
[34] Keith Vernon, 'Pus, Sewage, Beer and Milk: Microbiology in Britain, 1870–1940', *History of Science* xxviii (1990), p. 302.
[35] Waddington, 'Science of Cows', pp. 365–6.

Rather than representing overly routinised science that had to be fitted in around existing responsibilities, the institutional framework created on the farms and at Royalcot represented an experimental infrastructure that initially had few parallels. At the time, it was believed to offer the most sophisticated environment for examining the properties of bovine tuberculosis. More importantly, contemporaries hoped that it would be a 'valuable precedent for the State endowment of scientific research', meeting pressure from the science lobby for state support.[36] Through the infrastructure created, a considerable investment was made in defending the British model of bovine tuberculosis.

Science and the commissionn

Laboratory medicine 'meant an intimate admixture of pathology, bacteriology, physiology, chemical physiology, public and environment health' and this was exemplified in the scientific work of the royal commission.[37] However, from the start the commissioners conceived of bovine tuberculosis as a clinical problem, reducing the detection of the disease to the presence of the bacteria and a determination of its pathogenic properties. The investigators were essentially clinicians trained in a climate that maintained the value of clinical observation. Much of the work for the commission therefore initially followed similar lines to those pursued by other British and international researchers but on a larger scale than other contemporary investigations into bovine tuberculosis.[38] Considerable time and money went into confirming what the earlier commissions had proven as the commissioners sought to demolish Koch's claims. Earlier experiments were repeated and existing lines of inquiry were followed. Just as the questions the commissioners pursued were familiar, so too were the methods employed for cultivating and identifying pathogenic micro-organisms.[39] Transmission experiments, whereby samples of tissue or bacillus were taken from infected cows and then injected or fed to other cows or animals to determine levels of infection, had also come to be the cornerstone of testing for bovine tuberculosis and were almost routine. Yet, despite the traditional nature of many of the experiments, the commissioners

[36] *BMJ* ii (1907), p. 222; 'Royal Commission on Tuberculosis', *ibid.*, p. 331.

[37] David Smith and Malcolm Nicolson, 'The "Glasgow School" of Paton, Findlay and Cathcart: Conservative Thought in Chemical Physiology, Nutrition and Public Health', *Social Studies of Science* xix (1989), p. 215.

[38] For example, questions of virulence, and of morphological and chemical characteristics, which formed much of the focus for the commission's research, also represented an important strand in American tuberculosis research: see Georgina D. Feldberg, *Disease and Class: Tuberculosis and the Shaping of Modern North American Society* (New Brunswick, 1995), p. 130.

[39] For the experiments conducted by the commission, see Waddington, 'Science of Cows', pp. 368–9.

anticipated that these tried and tested methods would conclusively establish the zoonotic properties of tuberculosis where other researchers had been less forthright, pointing to the faith they placed in the nature of laboratory work. The commissioners hoped that in the process of these studies the different strains of tuberculosis and their cultural characteristics would be confirmed, their effect on animals measured, and virulence determined to clear up the identity of the disease and how dangerous different strains were. From the start, it was anticipated that this would then justify existing responses to the disease.

However, the commissioners did not intervene in the day-to-day research. Most were committed elsewhere and they were later to admit that they had given their researchers 'considerable latitude'.[40] This latitude allowed the commission to depart from its original remit but at the same time saw problems occur. Despite the expertise of the commissioners, the researchers effectively learned as they worked: experiments had to be duplicated to prevent error or false impressions, pointing not only to the continuing DIY nature of much of the work for the commission but also to the still rudimentary nature of bacteriological and pathological studies in the period. It took time for the researchers to gain the necessary experience and work out which methods to use. It quickly emerged that there were serious methodical flaws in the research conducted. Considerable anxiety was voiced in 1903 about the slapdash scientific approach of the investigators, which was hardly surprising given their relative lack of experience. For example, the method of conducting post-mortems was poor as was the record keeping.[41] Nor had an effort been made to count how many bacilli samples contained until Arthur Eastwood, the histologist in charge of the Royalcot laboratory, realised that this might create problems. After this, Eastwood would literally count the bacilli in cultures under a microscope – a process that involved 'severe eye strain' and could take anything between forty minutes and two hours. He admitted the rough nature of his estimates given the irregularity in how the bacilli were distributed.[42]

As the skills of the investigators gradually improved, they devised new methods. Although many of the experiments were not new, through working out these new methods the investigators helped refine bacteriology as a discipline. For example, given problems other researchers working in the field had experienced in producing suitable cultures, experiments were undertaken to determine the efficacy of using different media and in growing cultures artificially.[43] Although these experiments were incidental, they formed part

[40] *Royal Commission into the Relations of Human and Animal Tuberculosis*, PP (1907) xxxviii, p. 42.

[41] TNA: PRO, 'Memorandum on Post-Mortem Records', Dec. 1903, FD 22/7.

[42] TNA: PRO, Minutes of the Royal Commission on Tuberculosis, 19 Oct. 1903, FD 22/1.

[43] *Second Interim Report, Royal Commission into the Relations of Human and Animal Tuberculosis*, part ii, vol. ii, PP (1907) lvii, pp. 1–20, 21–6, 30.

of a learning process as problems were overcome. Difficulties with recognising the bacillus and in standardising the dose were also tackled by the investigators. This work in developing new ways of growing and identifying the bacillus was part of a general shift away from using animals as living cultures that had dominated pathology since the 1880s.[44] It also saw a refinement of bacteriological techniques that were adopted by other researchers.

Researching bovine tuberculosis

Despite the investment made in creating a research infrastructure, the scientific investigators were not working in a vacuum. Research into bovine tuberculosis continued to be conducted across the country, much of which was independent of the commission but followed similar lines of inquiry. Koch's pronouncement had produced a flurry of activity as doctors debated 'the many guises in which the disease might manifest itself'.[45] Attempts were made to find clinical cases to establish a clear link between bovine tuberculosis and the disease in humans. At the Mill Road Infirmary in Liverpool, for example, Nathan Raw conducted high profile studies into the morphological nature of bovine tuberculosis. His conclusions that Koch was right in asserting that there were two types of tuberculosis suggested that not all supported the British model.[46] Raw was, however, in a minority. In Aberdeen, David Hamilton, professor of pathology at the University of Aberdeen, and McLauchlan Young, lecturer on veterinary hygiene at Aberdeen, had taken up Koch's challenge to repeat his experiments, publishing their findings in 1903. They dismissed Koch's conclusions and argued that bovine tuberculosis was a zoonosis. Although their experiments were criticised, their work followed a similar methodological approach, addressed the same issues, and predicted the findings of the third commission. Young and Hamilton supported the common identity of bovine and human tuberculosis, though as with Raw, they noted that the 'morphological characters of the bacillus may vary' according to the host.[47] Their work, which was the first extensive study to be published in Britain following the 1901 congress, prompted considerable discussion. Other researchers repeated their claim and reinforced the idea that 'tuberculosis in all animals is generically one and the same disease' well in advance of the commissioners' report.[48]

[44] *Royal Commission Appointed to Inquire into the Effect of Food Derived from Tuberculous Animals on Human Health*, part iii, inquiry iii (London, 1895), pp. 172–3.
[45] 'A Discussion on the Relationship of Human and Bovine Tuberculosis', *BMJ* ii (1902), pp. 944–7.
[46] Nathan Raw, 'Human and Bovine Tuberculosis', *BMJ* i (1902), p. 237; *ibid.*, pp. 596–8.
[47] David Hamilton and McLauchlan Young, 'On the Relationship of Human Tuberculosis to that of Bovines', *Transactions of the Highland Agricultural Society of Scotland* (1903).
[48] Harold Swithinbank and George Newman, *Bacteriology of Milk* (London, 1903), p. 232.

Nor were investigations into bovine tuberculosis limited to Britain. The third commission has to be seen within an international context, although its work was by far the most extensive and widely cited. By the end of the nineteenth century, concerns about bovine tuberculosis were international in nature, and although no international eradication programme had emerged, many states employed similar methods to combat the disease. Just as in Britain, Koch's claims threatened these programmes, and researchers across Europe and the United States equally strove to confirm that Koch was wrong: too much had been invested in mechanisms to tackle bovine tuberculosis. Nor was Britain alone in launching state investigations. Commissions were established in Italy, Germany, Canada, Australia and Japan.[49] Much of the research addressed the same questions as the British commission. Although work in Italy and Japan played down the risks of bovine tuberculosis, the weight of international opinion went against the 'German *savants*'.[50] In France, for example, clinicians, bacteriologists and pathologists presented evidence that countered Koch's claims. Studies by Saturnin Arloing, professor of medicine in Lyon, noted that notwithstanding differences in character and virulence, 'the identity of human and bovine tuberculosis ought to be maintained, and the prophylactic measures which result from it ought also to be maintained'.[51] Work by Albert Calmette and others suggested that the tubercle bacilli could pass through the intestine and into the 'lymph stream, and from it into the circulation, to be filtered out by the lung, where they most commonly cause disease'. Calmette thought that this was 'the most common mode of infection'.[52] These views were confirmed by other European studies, many of which were widely reported in British medical and veterinary journals.

Similar views were expressed in the United States. Although 'talented and ambitious American medical men studied abroad in German universities' between 1870 and 1914 and hence were able to exert a powerful influence on American medical ideas, Koch's claims in 1901 generated a storm of protest.[53] Mazyck Ravenel, bacteriologist of the State Livestock Sanitary Board of Pennsylvania, led the attack against Koch at the 1901 conference and his Pennsylvanian laboratory continued to play a prominent role in defining bovine tuberculosis. Ravenel and the Bureau of Animal Industry repeatedly

[49] See Committee on Public Health, *Report of the International Commission on the Control of Bovine Tuberculosis, etc* (Ottawa, 1910); Advisory Council of Science and Industry Queensland Committee, *Report of Committee on Tuberculosis, with special reference to cattle and pigs* (Brisbane, 1917).

[50] 'Bovine and Human Tuberculosis', *BMJ* ii (1912), p. 1485.

[51] Saturnin Arloing, 'Experimental Demonstrations of the Unity of Tuberculosis', *JCPT* xvi (1903), pp. 138–51.

[52] Cited in Alfred Leffingwell, *American Meat, and its Influence upon the Public Health* (London, 1910), p. 52.

[53] Thomas N. Bonner, *American Doctors and German Universities: A Chapter in International Intellectual Relations, 1870–1914* (Lincoln, Nebraska, 1963), p. vii.

warned about the danger of bovine tuberculosis: they suggested that a substantial proportion of cases (even the pulmonary form) could be explained by the ingestion of tubercle bacilli.[54] Ravenel expressed a common sentiment when he noted that, experimental evidence aside, 'one cannot study such statistics as those given without being fully convinced that a very important proportion of the children who die of tuberculosis are infected through their food'. For him, as for many researchers, bovine tuberculosis was 'a menace to human health'.[55]

In making these claims, the need to continue to target meat and milk was defended at an international level. Koch was accused of taking 'undue advantage' of his position as the tide of scientific opinion turned against him.[56] Even in Germany, where support for Koch took on an official tone, dissent was voiced. For example, Robert Ostertag, in his authoritative *Handbook of Meat Inspection*, suggested that Koch's evidence was insufficient. He explained that 'all organs affected with tuberculosis must . . . be excluded from the market as dangerous food materials'.[57]

Koch continued to defend his position, arguing that 'in our efforts to combat tuberculosis it is important that we should not lose ourselves in wrong paths if success is to be achieved'. For Koch the considerable international effort directed at bovine tuberculosis constituted such a wrong path, and for him this ensured that the much greater threat from the consumptive patient was diluted.[58] The German investigation did confirm his views but it did not reflect a consensus as Koch's claims were challenged by some of his countrymen. The stance of the German investigation was hardly surprising. Although it took decades of debate to reach an agreement on what bacteriology and germ theory meant, Koch was lionised in Germany and became a source of national pride. He was considered to 'embody the transition to a new scientific epoch in medicine' and his initial pronouncement on tuberculosis to the Berlin Physiological Society was regarded as 'the watershed moment', heralding the start of this era.[59] By the 1910s, however, Koch was being forced to retreat, even if his professional position remained unassailed.

[54] Leffingwell, *American Meat*, pp. 49–52.
[55] Mazyck Ravenel, 'The Intercommunicability of Human and Bovine Tuberculosis', *JCPT* xv (1902), pp. 140, 143.
[56] Cited in Barbara G. Rosenkrantz, 'The Trouble with Bovine Tuberculosis', *BHM* lix (1985), p. 157.
[57] Ostertag, *Handbook*, pp. 629–31.
[58] Robert Koch, 'The Transmissibility of Bovine Tuberculosis to Man', *JCPT* xv (1902), p. 299.
[59] George E. Haddad, 'Medicine and the Culture of Commemoration: Representing Robert Koch's Discovery of the Tubercle Bacillus', *Orisis* xiv (1999), p. 118.

Confirming the model

As more international studies were published that pointed to the 'fallacy of Professor Koch's reckless statement', the British commission came under pressure to deliver a report just as anxiety for more effective measures to promote a pure milk supply was building.[60] The LGB, despite its reputation as obstructive of social reform through ideological conservatism and administrative incompetence, wanted to see results. By the Edwardian period, the LGB had become more active when it came to the issue of food safety: it had already appointed inquiries into food safety and was under pressure to provide local authorities with more information about food production and public health.[61] Food was also moving up the political agenda as part of the Liberal government's goal of promising cheaper food by reducing import duties. Political leverage was applied on the LGB by MPs with an interest in food regulation. At the same time, the LGB was also under pressure from the meat trade, which wanted to see compensation introduced when tuberculous meat was seized, and from local authorities, which wanted to know the results of the commission to support their efforts at further regulation.[62] Despite prompting from the LGB, the commissioners initially stalled, arguing that conclusive findings required 'a very wide series of experiments', but under pressure it was decided to issue a short interim report.[63]

The resulting report was hastily put together to answer external pressure rather than to reflect firm conclusions. As such it was superficial. The report briefly stated that the human tubercle bacilli could not be converted into the bovine tubercle bacilli when passed through cows and that 'human and bovine tubercle bacilli' were not 'specifically different'. This was a more cautious approach than that suggested by the investigators, who felt that the two diseases were 'identical' as 'regards their bacteriological characteristics'. The conclusions were the same however: the commissioners were clear that there was no need to alter existing legislation.[64]

Despite its support for the maintenance of the status quo, the report was a considerable disappointment. It bolstered up 'the accepted English medical theory' but did little to provide the detailed evidence or promote the action that many were looking for to counter Koch's claims given his iconic status in

60 'Human and Bovine Tuberculosis', *Public Health* xv (1902/3), p. 685.

61 Kenneth D. Brown, 'John Burns at the Local Government Board: A Reassessment', *Journal of Social Policy* vi (1977), pp. 157–70; Michael French and Jim Phillips, *Cheated Not Poisoned? Food Regulation in the United Kingdom, 1875–1938* (Manchester, 2000), pp. 66–95.

62 LMA: LCC, *Report of the Public Health Committee for the Year 1902* (1903), p. iv.

63 TNA: PRO, Minutes of the Royal Commission on Tuberculosis, 7 Mar. 1904, FD 22/2.

64 TNA: PRO, 'Modification Experiments', FD 22/9; *First Interim Report, Royal Commission into the Relations of Human and Animal Tuberculosis*, PP (1904) xxxix, p. 6.

bacteriology.[65] Given Young and Hamilton's work, it is hardly surprising that the findings of the commission met with dissatisfaction. The notion that bacteria and other micro-organisms could change their character and properties, depending on their environment, was already accepted, and the idea that the bovine and the human bacilli were morphologically different was no longer a matter of debate.[66] In addition, the report left many questions unanswered, particularly about the exact identity of bovine and human tuberculosis and the relationship between them.

This confusion encouraged the commissioners to produce a further set of memoranda in 1905 and an interim report in 1907 in response to Koch's 1906 Nobel Lecture in which he reasserted that human and bovine tuberculosis were distinct diseases. This time the commissioners offered detailed experimental evidence as a direct challenge to Koch and his supporters. Although the second report repeated earlier assertions and reflected claims made by other researchers, according to Dwork, it 'conclusively clarified what had come to be understood prior to the British Congress on Tuberculosis in 1901'.[67] Convinced that the inoculation experiments had produced 'identical' tubercular lesions no matter which bacilli were used, the commissioners asserted that bovine tuberculosis 'invariably produces acute tuberculosis' in other animals and that 'bovine animals and man can be reciprocally infected'. For the commissioners, 'there seems no valid reason for doubting the opinion . . . that human and bovine tubercle belong to the same family'.[68] Morphological differences were acknowledged and although it was felt that the human bacillus produced a less acute disease in domestic animals, the main distinction was seen to be one of virulence not type. These conclusions overlooked the possibility that subtle differences existed between the bacilli, which were dismissed by Eastwood as a product of artificial cultures. In putting forward these views, the commissioners confirmed existing notions that virulence was the key and asserted a mechanical view of infection based on 'blood vessels and lymphatic channels'. Based on Eastwood's histological work, it was asserted that the tubercle bacillus could be present in organs and in the blood in large numbers without causing lesions.[69] This was a tacit endorsement of a generalised model that allowed for more rigorous control of the meat trade. In adopting this stance, the commission served to justify the stance adopted by MOsH in their attempts to condemn tuberculous meat.

The second report was better received largely because it presented the comprehensive evidence the medical and scientific community wanted. The

[65] 'Tuberculosis Commission Report', *Cowkeeper and Dairyman's Journal*, June 1904, p. 297.
[66] See *PRSM* iii (1910), p. 218.
[67] Debora Dwork, *War is Good for Babies and Other Young Children: A History of the Infant and Child Welfare Movement in England, 1898–1918* (London, 1987), p. 83.
[68] *Second Interim Report*, pp. 13, 68–70; *ibid.*, part ii, appendix, vol. iv, p. xxx.
[69] *Ibid.*, part ii, appendix, vol. iii, pp. 221–4.

BMJ felt that the report 'dealt a death blow to a doctrine which, during the last few years, has exercised a mischievous influence upon the interests of public health'.[70] For *Public Health*, the report 'proves, once and for all, that a certain number of cases of human tuberculosis, especially of tuberculosis in children, are the direct result of the introduction into the human body of the bacillus of bovine tuberculosis'. This legitimised existing attempts to control the disease. The journal anticipated that local authorities would use its findings to step up their efforts to combat bovine tuberculosis, as it was in prevention that 'our chief means for diminishing tuberculosis' was believed to exist.[71] The publication of the report increased demand for legislative action, and its findings were used by medical and veterinary officers at a local level to support their calls for reform.

Koch was not so optimistic. Opposition to his views at international conferences since 1901 had made him dig his heels in and impute personal rather than professional motives to his opponents.[72] He dismissed the British report at the 1908 Washington congress on tuberculosis. A special forum on bovine tuberculosis had been organised at the congress, although it is unclear whether this was at the request of Koch or Ravenel. Koch certainly felt his reputation was at stake and therefore argued that the need to distinguish between the bacilli was only theoretically interesting. He continued to insist that attention should focus on pulmonary tuberculosis; that bovine tuberculosis was a negligible factor in human tuberculosis. Although he gave Theobald Smith credit for drawing attention to the differences between the bovine and human bacilli, Koch attacked the experimental methods used by the British commissioners. There was some truth to Koch's accusations, which reflected growing international efforts to combat the pulmonary form of the disease. However, delegates to the congress were not convinced. After hearing evidence from Woodhead and MacFadyean, and from other researchers who confirmed the findings of the British commission, they unanimously pressed for measures to prevent the spread of bovine tuberculosis to humans.[73]

The differences between the commission and Koch excited further public attention. Both the press and the LGB rallied behind the commissioners out of national pride for the science involved. For the *BMJ*, 'Koch has lost the battle and now stands almost alone in the field.'[74] Given anxieties about the relative positions of German and British science, these were important claims to make. Under attack and accused of inconsistencies in his work, Koch retreated. By 1909, in an interview with *The Times*, he adopted a less disparaging view, claiming 'he had no desire to cast an aspersion upon British

[70] *BMJ* i (1907), pp. 330–2.
[71] 'Royal Commission on Tuberculosis', *Public Health* xix (1906/7), pp. 469, 471.
[72] See 'Koch and Tuberculosis', *Veterinary Record*, 15 Nov. 1902, p. 320.
[73] 'Bovine and Human Tuberculosis', *Sanitary Record*, 18 Mar. 1909, p. 234.
[74] *BMJ* ii (1908), p. 1201.

methods'.[75] Three years later he admitted that the only results that should be considered of value were his and those obtained by the British commission.[76] In the following year, the German Imperial Health Commission officially departed from Koch's initial stance.[77] It was with Koch's retreat that the question of bovine tuberculosis appeared to be settled in favour of the British model.

The final report, issued in 1911, asserted the importance of 'experimental method'. In doing so, it confirmed the value of bacteriological and pathological studies over epidemiology in the investigation of disease that had an impact beyond the commission's findings. The commissioner admitted that without this research 'we should not have been able to give any answers to the questions referred to us'.[78] Yet, despite the endorsement the commissioners gave to experimental science at a time when neglect of laboratory studies in England was being attacked, the final report said little that was new. It repeated accepted ideas about the nature, characteristics and uniformity of the bacillus, along with how it spread through the body, and identified three forms of bacillus – human, bovine and avian – confirming findings first made in 1891. The commissioners reasserted that bovine tuberculosis could cross the species barrier and argued that they had 'conclusively shown that many cases of fatal tuberculosis in the human subject have been produced by the bacillus known to cause the disease in cattle'. This appeared to offer the certainty that many had been looking for since the 1889 Glasgow trial. Based on this evidence the commissioners argued that existing regulations on meat and milk should 'not be relaxed' but enforced 'to afford a more adequate safeguard against the infection of human beings through the medium of articles of diet from tuberculous animals'.[79] It had taken ten years and a considerable investment to confirm what many already felt to be true.

Framework for research

The commissioners realised that, despite ten years of research, many questions remained unanswered. However, they were also aware that the royal commission could not sit indefinitely. In 1908, they had told the LGB that 'some other similar establishment' to the laboratories at Stansted would be needed, probably under the funding of the LGB, to investigate important 'Public Health' issues.[80] For the commissioners their work had demonstrated the

[75] Cited in *Veterinary Record*, 9 Jan. 1909, p. 460.
[76] *Veterinary Journal* lxviii (1912), p. 550.
[77] 'Bovine and Human Tuberculosis', *BMJ* i (1913), p. 845.
[78] *Final Report, Royal Commission into the Relations of Human and Animal Tuberculosis*, p. 3.
[79] *Ibid*. pp. 4–29, 35, 36–7, 40–1.
[80] TNA: PRO, Minutes of the Royal Commission on Tuberculosis, 2 Nov. 1908, FD 22/3.

importance of a national research institution and the need to move beyond a style of scientific investigation based upon using existing facilities and researchers in other institutions established in the mid nineteenth century under John Simon. Keen for research into tuberculosis to continue, they told the Departmental Committee on Tuberculosis set up following the 1911 National Insurance Act that the small research fund included in the Act would facilitate the research they had initiated.[81] Government funding for large-scale national and systematic schemes for scientific research was beginning to be developed and the commissioners' recommendations should be understood in this context. The commissioners also joined other eminent medical authorities in issuing a memorandum to the Departmental Committee arguing that the fund should be 'applied definitely and specifically' to research into tuberculosis.[82]

The Medical Research Committee (MRC), set up by the fund, continued to support research into tuberculosis and in the interwar period funded appointments for many of the commission's researchers. For Vernon, the work of the royal commission was important in shaping the creation of the MRC, whilst the commissioners' conception of a research institute prefigured the MRC's efforts to create dedicated research centres. However, whilst the MRC maintained an interest in bovine tuberculosis convinced that the disease 'is a matter of great moment in public health owing to the bovine source of much tuberculosis infection among human beings', the physical structure of the royal commission was taken over by the LGB.[83] Eastwood and Griffith transferred much of the research equipment from Stansted to the LGB's new laboratory in Carlisle Place so that research into bovine tuberculosis could continue. Eastwood and Griffith followed the equipment and continued to work in the field, but as attention increasingly shifted in the interwar period to questions of eradicating tuberculosis in cattle and treatments for the pulmonary form of the disease, the framework established by the royal commission became less important.[84]

The model defended

After the débâcle with Koch in the aftermath of the Washington congress, the conclusions put forward by the commission had already been accepted. The

[81] Oxford University, Bodleian Library: Evidence, 1912–13 Departmental Committee on Tuberculosis, MSS Addison, dep. C. 8.
[82] Oxford University, Bodleian Library: Memorandum from medical schools, MSS Addison, dep. C. 15.
[83] TNA: PRO, Fletcher to Treasury, 23 July 1923, T 161/213.
[84] TNA: PRO, Minutes of the Royal Commission on Tuberculosis, 4 Apr. 1910, 25 July 1910, FD 22/3.

commissioners, if they voiced 'little which has not been known to the world's scientists for some years', did, however, 'put a seal of finality on certain theories; the seal of falsity on others; which hitherto', according to the *Sanitary Record*, had 'struggled for the mastery, and . . . the applause of the public'.[85] As one veterinary inspector explained, 'to-day the doubt raised by Koch as to the communicability of bovine tuberculosis to humans has been completely dispelled'.[86] The mass of evidence collected by the commissioners confirmed the British model and what other researchers, policy makers, and scientists felt they already knew. As such, the evidence presented was used to support existing ideas, whilst the methods developed by the investigators were incorporated into bacteriological practices.[87] On the back of the report, local authorities started to draw up bills to tighten their systems of meat and milk inspection and to promote more effective control. In many respects, the commissioners' report appeared to settle the questions that had surrounded bovine tuberculosis since the 1870s. In doing so, it justified action against the disease and legitimised the public health policies already adopted.

[85] 'Royal Commission on Tuberculosis', *Sanitary Record*, 20 July 1911, p. 63.
[86] GCA: Report of the Veterinary Surgeon to the Corporation of the City of Glasgow, 1911–12, C2/1/12.
[87] See Arthur Newsholme, *The Prevention of Tuberculosis* (London, 1908).

8

Diseased meat control and the 'great modern scourge'

According to the *Veterinary Record* by 1899 bovine tuberculosis had become 'the burning question of the day'.[1] Although the journal exaggerated the importance of the disease as part of its campaign to secure greater influence for veterinarians in the control of meat and milk, statistical evidence that infection through the digestive tract was common heightened fears and contributed to a sense that bovine tuberculosis was an 'immense danger'.[2] Anxiety was further fuelled by a growing interest in food safety encouraged by a decline in the major infectious diseases and a rise in food-borne infections. If by the Edwardian period many issues regarding food and contagion remained to be authoritatively settled, questions about food safety and food-borne infections had become important public health concerns as attention focused on specific disease agents. Bovine tuberculosis stood at the heart of these anxieties and concerns about zoonotic disease. Although attention was deflected away from the disease by the sanatorium movement and an awareness that most cases of tuberculosis were caused by human-to-human transmission, many doctors and veterinarians could agree with the MOH for Cardiff that efforts to protect the public from bovine tuberculosis were key to 'a general scheme for the prevention of tuberculosis'.[3] It was argued that because the consumer exercised little discretion, and because bovine tuberculosis could infect humans, it was essential to have a system of meat inspection that would protect the public. At a local level, further attempts were made in the face of opposition from the meat industry to improve inspection and demand for statutory powers to target known sources of tuberculous infection in food increased. Although faith continued to be placed in the value of inspection to prevent the sale of tuberculous meat, by the Edwardian period many had become convinced that the only solution lay in eradication.

[1] *Veterinary Record*, 9 Sept. 1899, p. 143.
[2] *Bibby's Book on Milk*. Section IV. *Bovine Tuberculosis: Cause, Cure and Eradication* (Liverpool, [1911]), p. 8.
[3] GRO: Special health committee, 30 May 1899, BC/C/6/19.

An expanding service?

Fears about bovine tuberculosis continued to underpin efforts to improve meat inspection and faith continued to be placed in the value of systematic inspection and in the mechanisms established under the 1875 Public Health Act to protect the consumer from 'any communicable disease', parasites and 'from buying inferior meat'.[4] Surveillance and seizure remained key. However, despite efforts to extend inspection, Britain lagged behind its European counterparts. From the 1890s onwards, many European states moved to establish uniform systems of meat inspection. In Britain, this uniformity was frustrated by an impasse over compensation, by strong local traditions, and by the powerful meat lobby, which blocked moves to establish compulsory controls and ensured that initiatives continued to be shaped by local needs. The low status of sanitary inspectors and veterinarians prevented concerted attempts to improve the regulation of the urban meat supply, as they struggled for professional recognition in other areas of their work. By the outbreak of the First World War, many of the problems that had troubled the second royal commission on tuberculosis continued to frustrate inspection, despite efforts by sanitary authorities and anti-tuberculosis organisations to increase the vigilance of inspection.

Although progress in Britain did not match European efforts, the first and second commissions on tuberculosis did stimulate moves to improve inspection. In other areas of public health work, new officers were being appointed and services were being expanded as a community of interests surrounding preventive medicine developed. Moves to extend inspection was part of this growth, but mounting public interest in food purity also exerted pressure on local authorities to act. Although most of the work remained the responsibility of overworked sanitary inspectors, many MOsH sought to persuade their sanitary committees to employ special officers to inspect markets and slaughterhouses. In Holborn, for example, after several years of protests from William Bond the MOH, the Board of Works decided to investigate the meat trade in the district. Interest followed the Glasgow trial and coincided with a burst of activity in the Board and the appointment of new officers. It was quickly found that inadequate inspection had allowed Holborn to become a haven for the sale of diseased meat. In response, the Board appointed G. Billing as meat inspector in 1895. There was an immediate improvement: under Billing, 'large quantities of diseased meat were "seized" and condemned' in the first two months, although thereafter the number of cases fell.[5] Similar moves to appoint

[4] Arthur Littlejohn, *Meat and its Inspection: A Practical Guide for Meat Inspectors, Students, and Medical Officers of Health* (London, 1911), p. 1.

[5] Holborn District, *Report of the Medical Officer of Health for the half-year Ending December 1895* (London, 1896), pp. 19–20; 'Holborn Meat Market', *Public Health* viii (1895/6), p. 385.

meat inspectors were repeated elsewhere and nationally the amount of diseased meat seized increased.[6]

Attention continued to be directed at raising the quality of the inspectors as part of efforts to professionalise public health work. The creation of a body of experts was cast as an important mechanism to increase the chance of detection and a solution to some of the criticisms that had been levelled at meat inspection practices. It was argued that 'so long as we continue to work by rule of thumb' and allow inspection to be conducted by untrained men 'we may expect to get scant, or too full, measure of convictions'.[7] The need for qualified meat inspectors was reinforced by the second Royal Commission. Alarmed by the chaotic system of inspection, it recommended that 'no person be permitted to act as a meat inspector until he has passed a qualifying examination' and was conversant with the 'signs of health and disease in animals destined for food', and the 'appearance and character' of meat.[8] The Sanitary Inspectors Association, conscious of the growing importance ascribed to meat inspection as part of the day-to-day work of a sanitary inspector, added its voice in a bid to raise the status of sanitary inspectors. The LGB therefore applied pressure on local authorities to appoint a qualified 'separate officer for the purpose'.[9]

The publicity generated by the alarm surrounding tuberculous meat, and demands for qualified inspectors, reinforced an existing desire to establish a national qualification to improve the nature and quality of sanitary inspection. Although the Sanitary Inspectors' Examination Board was finally set up in 1897 after six years of wrangling, problems from the outset ensured that a patchwork of qualifications existed. Competing courses were established that reflected the different conceptions of sanitary inspectors' duties. However, all paid attention to 'diseases of animals rendering Flesh unfit for Consumption'.[10] Sanitary inspectors were aware that meat inspection required 'special knowledge', skill and judgement.[11]

The beleaguered Sanitary Institute set up the most comprehensive course in meat and food inspection in 1896; a separate examination for meat inspectors followed three years later. The course was designed to equip students with a thorough knowledge of the laws governing food inspection, and ensure familiarity with the signs of diseased and healthy animals 'when alive and after

[6] See GRO: Cardiff health and port sanitary committee, 1 Feb. 1897, 23 Feb. 1897, BC/C/6/16.

[7] 'Tuberculous Meat', *BMJ* i (1891), p. 419.

[8] *Royal Commission Appointed to Inquire into the Administrative Proceedings for Controlling the Danger to Man through the use as Food of the Meat and Milk of Tuberculosis Animals* (London, 1898), p. 7.

[9] Francis Vacher, *The Food Inspector's Handbook* (London, 1913), p. 2.

[10] TNA: PRO, Sanitary Inspectors' Examining Board, MH 26/1; TNA: PRO, Allan and James to LGB, 24 Feb. 1893, MH 26/1.

[11] 'Practical Meat Inspection', *Sanitary Inspectors' Journal* ii (1895–6), p. 392.

slaughter'. Despite accusations that the Institute's examination was too theoretical, a complaint that was also applied to the DPH, practical demonstrations were offered at the London meat market at West Smithfield, and in 1903 further courses were established in Leeds, Bradford and Huddersfield. In addition to attending the course, candidates were expected to present evidence of two months' experience at a cattle market or abattoir.[12] It was anticipated that the examination would go some way to addressing calls from butchers for trained inspectors and suggestions were made that attendance should be compulsory 'as that subject forms an integral part of a Sanitary Inspector's duty'.[13] Although the course was attacked by veterinarians, anxious that it would undermine their claims to be the main arbiters of diseased meat, within the first year 300 students had attended. By 1925, the Institute had examined 3,039 candidates and issued 1,665 certificates.[14]

Pressure was applied on those already in post to sit the Sanitary Institute's examination, but most sanitary and meat inspectors, given the full-time nature of their work and lack of practical support, did not take advantage. Inspectors were appointed on the condition that they pass the examination, but even when they failed to do so their appointments were invariably renewed. The *Medical Officer* reported that 'in the large majority of instances' those responsible for inspecting meat 'had no special training'.[15] Meat inspection continued to be tacked on to sanitary inspectors' general duties, in part because many local authorities could not afford to make separate appointments. This had a marked effect on the growth of meat inspection as a specialism and contributed to existing criticisms of the quality of inspection. George Reid, writing in the *Sanitary Inspectors Journal*, felt that the low status of the posts discouraged local authorities from requiring qualifications, whilst sanitary authorities bemoaned the difficulties of employing inspectors 'with sufficient practical and theoretical knowledge'.[16] One contemporary commented that 'looking at the list of trades represented amongst meat inspectors one would think that the local authorities vied with each other in trying to find the most unsuitable men for the purpose'.[17] In Portsmouth, for example, three inspectors had formerly been teachers, one a dispenser, one a carpenter and another a tram conductor. Battersea relied on four plumbers and three carpenters.[18] Most continued to

[12] TNA: PRO, Coles to Owen, 23 Mar. 1893, MH 26/1.

[13] 'Practical Meat Inspection', *Sanitary Record*, 26 Mar. 1908, p. 270.

[14] 'Inspection of Meat', *Food and Health*, 28 May 1897, p. 59; Louis C. Parkes, *Jubilee Retrospect of the Royal Sanitary Institute, 1876–1926* (London, 1926), p. 34.

[15] 'Veterinary Public Health Officers', *Medical Officer*, 23 Jan. 1909, p. 569.

[16] George Reid, 'The Position of Sanitary Inspections', *Sanitary Inspectors' Journal* ii (1895–6), p. 258; CLRO: Report on meat inspection, Nov. 1906, PH/159.

[17] H. Bowes, 'Diseases of Domesticated Animals Communicable to Man', *Veterinary Record*, 19 May 1900, p. 667.

[18] *Royal Commission Appointed to Inquire into the Effect of Food Derived from Tuberculous Animals on Human Health* (London, 1895), p. 7.

be butchers by trade, however. In the City of London, for example, the eight meat inspectors were all 'men who have had practical experience as butchers'. Such men were felt to make 'excellent inspectors'.[19] Good practical knowledge and experience were considered better than training.

Although local authorities continued to appoint unqualified men as sanitary inspectors or butchers as meat inspectors, a growing body of literature was produced to aid them in their work. The science and deductive art of inspection had improved considerably since the manuals of the 1880s. Handbooks for sanitary inspectors preparing for the Sanitary Institute's examinations were in demand and sections in public health manuals devoted to food inspection grew dramatically.[20] Many drew heavily on translations of Robert Ostertag's authoritative *Handbook of Meat Inspection*, with a large proportion written by those MOsH involved in campaigns to improve the urban meat trade.[21] Concerns about bovine tuberculosis lay behind attempts to systematise a body of knowledge that was felt to be essential for meat inspectors and the disease was discussed in detail because it was 'so important from a public health point of view'.[22] Practical guidance remained key. Manuals explained the law relating to meat inspection, slaughtering practices and the condition of normal meat, but paid most attention to diseased and tuberculous meat. Suggestions were made on how to make accurate deductions about the state of carcasses. As with earlier guides, manuals detailed the 'unnatural appearances' that inspectors should look for in carcasses. Inspectors were advised to make themselves 'thoroughly acquainted with the diseases of animals which render the meat unfit' and were told to be systematic and to spend time examining the organs.[23] However, not only did manuals seek to supply 'an aid to those who have to deal practically with the subject', they also emphasised the value of tact, sympathy amd fairness.[24] These attributes were an important part of the professionalisation of sanitary inspection and represented a gentlemanly ideal that reflected a rhetoric common in public health appointments.

Inspectors and MOsH could also supplement these predominantly visual examinations with bacteriological tests to determine the presence of the tubercle bacillus, especially in doubtful cases. Following the appointment of more bacteriologically aware MOsH in the 1890s, sanitary officials made greater use of laboratory tests to identify disease agents as pleas were made for further bacteriological studies into food-borne infections. Bacteriology allowed MOsH to assert their scientific credentials to counter opposition from those

[19] CLRO: Public health committee, 2 May 1905; Vacher, *Handbook*, p. 3.
[20] See John Cowderoy, *Notes on Meat and Food Inspection for Sanitary Inspectors* (Kidderminster, 1912); A. E. Bonham, A *Practical Guide to the Inspection of Meat and Foods* (Exeter, 1915).
[21] Robert Ostertag, *Handbook of Meat Inspection* (London, 1904).
[22] Littlejohn, *Meat*, p. 168.
[23] Albert Taylor, *The Sanitary Inspector's Handbook* (London, 1906), p. 282.
[24] Littlejohn, *Meat*, p. v.

who resisted prosecution or rejected the need for action. As *Public Health* explained, although 'bacteriological diagnosis is as yet by no means perfect' it still 'affords a very considerable measure of assistance in the struggle with infectious disease'.[25] Part of a trend towards a greater reliance on bacteriological tests in public health work, which helped redirect and improve the efficiency of public health medicine after 1890, the investigation of food became an important component of the work undertaken by the growing number of laboratories set up in the 1890s for public health work. From its creation in 1899, the Cardiff public health laboratory, for example, received 'a large number of specimens of diseased meat' for analysis. Its work was 'invaluable in connection with the examination of meat' and meat seizures in Cardiff came to depend 'in each instance upon the results of the examination'.[26] There were limits, however. Bacteriological studies could not determine how safe meat was: as John Washbourn, professor of bacteriology at Guy's Hospital noted, an absence of obvious signs of the tubercle 'does not absolutely preclude tubercle, for in some cases, especially when caseation has not occurred, the bacilli may not be present in sufficient numbers to detect'.[27] In addition, bacteriological procedures remained expensive and time-consuming. Most sanitary authorities therefore continued to rely on the authority of magistrates to determine whether meat was diseased or not.

By 1914, those involved in meat inspection could point to a standard examination and a wealth of advice literature on how to inspect meat. The number of inspectors had increased and most local authorities had a rudimentary system of meat inspection, even if the competency of those responsible remained open to question. However, uncertainty remained about what constituted diseased meat. Nor could contemporaries decide who was best qualified to undertake inspection. These tensions were revealed over the appointment of veterinary inspectors.

'Any other person but the veterinary surgeon'

Although meat inspection remained the responsibility of overworked sanitary inspectors and (in urban areas) specially appointed inspectors, a growing number of local authorities started to make greater use of veterinary surgeons. In Europe, the need for qualified veterinary inspectors had already been accepted as an important component of meat inspection in the late 1880s. However, Britain was slower to embrace the idea that veterinarians had a central role to play. It was only from the 1890s that support outside of the

[25] 'Bacteriological Laboratories', *Public Health* xv (1903), p. 179.
[26] GRO: Cardiff health and port sanitary committee, 23 Dec. 1902, BC/C/16/2; *ibid.*, 11 Mar. 1902, BC/C/6/24.
[27] CLRO: Public health committee, 22 Nov. 1898.

veterinary profession started to be expressed. Witnesses to the royal commission of 1896 to 1898 argued that veterinary surgeons should be responsible for meat inspection and the commissioners came to similar conclusions.[28] Individual doctors came out in support of the need for veterinary health officers to inspect all animals intended for food both before and after slaughter. Emphasis was placed on the need for 'a partial veterinary training' as 'an essential qualification for the suggested new class of food inspectors'.[29]

Although veterinary surgeons had been among the first to point to the dangers of bovine tuberculosis, asserting their expertise in identifying the disease in both living cattle and dead meat, they were assigned a marginal role when it came to inspecting meat. Frank Garnett, a veterinary surgeon in Windermere, voiced a common complaint when he wrote in the *Veterinary Record* that it was laughable that 'the invariable answer' to the 'fitted person' to determine the safety of meat is 'the MOH, butcher, shoemaker or any other person but the veterinary surgeon'.[30] The low status of veterinary medicine and its associations with trade ensured that they were often dismissed as unscientific practitioners and their views regarded as being little better than quackery. The debate about the contagious properties of bovine tuberculosis, however, underlined the need for qualified practitioners to undertake the inspection of meat and animals. In response, veterinarians argued that they should play a central role in meat inspection, calls that became caught up with veterinarians' professional demands for greater recognition.[31] They claimed an ontological expertise, and in campaigning for meat inspection to be carried out by qualified practitioners they asserted their credentials to be appointed inspectors. For the newly formed Association of Veterinary Officers of Health, the active involvement of veterinarians was essential to protect the public and it argued that only a comprehensive inspection by a trained veterinary surgeon could stop the sale of tuberculous meat.[32] Yet claims that veterinarians were 'specially trained and examined in meat inspection' were not entirely accurate. Veterinary training was poor throughout the nineteenth century and had low entry standards.[33] In addition, meat inspection had only been included in the revised four-year curriculum introduced in 1895 following the report of the

[28] *Royal Commission . . . into . . . Controlling the Danger to Man through the use as Food of the Meat and Milk of Tuberculosis Animals*, minutes of evidence, pp. 70, 161; *ibid.*, part i, p. 8.

[29] 'The Education of the Meat Inspector', *Veterinary Journal*, July 1906, p. 46.

[30] *Veterinary Record*, 25 Dec. 1897, p. 360.

[31] For a discussion of veterinary involvement in public health, see Anne Hardy, 'Professional Advantage and Public Health: British Veterinarians and State Veterinary Services, 1865–1939', *Twentieth Century British History* xiv (2003), pp. 1–23.

[32] 'Veterinary Health Officers and Inspection of Meat', *Sanitary Record*, 10 Sept. 1908, p. 241.

[33] See John R. Fisher, 'Not Quite a Profession: The Aspirations of Veterinary Surgeons in England in the Mid-Nineteenth Century', *Historical Research* lxvi (1993), pp. 290–1.

first royal commission on tuberculosis. As one veterinarian reluctantly commented, 'the vast majority of our members' were 'without systematic training in the subject at all'.[34] Nor were they trained in all aspects of food inspection, making them ill-suited to the multiple duties local authorities tended to expect of their officers.

Veterinarians were inclined to overlook these problems because of the professional and financial advantages of becoming involved in public health work. The profession in the mid nineteenth century had 'minimal market penetration' outside of the equine sector and a high degree of competition from unqualified practitioners.[35] A scientific knowledge of animal physiology and a commitment to contagionism did little to improve their status, especially when many veterinarians were not necessarily more competent than those old-style animal doctors they were meant to supplant. With most veterinarians in a precarious financial and professional position, involvement in public health was seen as a means of improving their status and security in much the same way as doctors had benefited from being brought within the public health apparatus in the 1840s. Meat inspection offered one route to achieve this, despite it being considered 'dirty work' and beneath professional men. Competition for veterinarian posts could therefore be intense: for example, when Glasgow advertised for a veterinary inspector, sixty-two applications were received.[36] Increasingly, the cow and tuberculosis were seen as a 'professional lifeline' and as a way of legitimising veterinarians' expertise and standing.[37]

Yet, despite growing recognition that diseased animals posed a threat to human health, only a few sanitary authorities employed veterinary surgeons as nuisance or meat inspectors. Even when veterinary surgeons were appointed, most of their work at first had little to do with meat inspection. Most were employed on a part-time basis and were concerned with the care of the corporation's horses and the monitoring and control of epizootics.[38] Only occasionally were these veterinary officers called upon to give their opinion about diseased carcasses.[39] The 1878 Contagious Diseases (Animals) Act did lead to the appointment of more veterinary inspectors and, as local authorities moved to prosecute those caught selling tuberculous meat, their duties expanded to include meat inspection. In Glasgow, responsibility for meat inspection was transferred in 1897 from the dual control of the police and sanitary inspector to Alexander Trotter, veterinary surgeon and later zealous

[34] 'Meat Inspection', *Veterinary Record*, 8 Dec. 1906, p. 368.

[35] Fisher, 'Not Quite a Profession', p. 293.

[36] GCA: Health committee, 4 Jan. 1897, E1/20/17.

[37] Hardy, 'Professional Advantage and Public Health', p. 10.

[38] William Hunting, 'Address to Conference of Veterinary Inspectors', *Journal of the Sanitary Institute* xx (1899), p. 446; Anne Hardy, 'Pioneers in the Victorian Provinces: Veterinarians, Public Health and the Urban Animal Economy', *Urban History* xxix (2002), pp. 382–4.

[39] See 'Tuberculosis at West Hartlepool', *Veterinary Record*, 21 Sept. 1889, pp. 158–9.

secretary of the Association of Veterinary Officers of Health. The appointment was designed to allow James Russell, the MOH, to concentrate on the prevention of pulmonary tuberculosis and end the fragmented nature of meat inspection that had characterised the city. Trotter was appointed to examine cattle 'with a view to the detection of Tuberculosis'.[40] By 1901, a veterinary section had been set up: it had a staff of seven qualified meat inspectors and an assistant veterinary surgeon whose efforts were mainly directed at seizing tuberculous meat. Under Trotter, meat inspection in Glasgow became 'more systematic'.[41] Other local authorities followed: Birmingham, Liverpool and Sheffield had all appointed veterinary meat inspectors by 1910.[42] Elsewhere action was more modest and built on existing provision under the Contagious Diseases (Animals) Acts. In Edinburgh and Swansea, for example, the duties of the veterinary inspector appointed under the Acts were extended to cover the sale of meat, the latter provoking opposition from the Swansea Butchers' Association who claimed a 'malicious' assault on their working practices as the number of prosecutions increased.[43]

Although growing enthusiasm was expressed in the need to employ veterinary officers of health, appointments remained patchy. Sanitary authorities were wary of the cost implications of employing qualified veterinarians who were more expensive than relying on the existing system of inspection by sanitary officials. Veterinarians recognised that MOsH, despite the rhetoric, were unwilling to give up what they saw as 'a sort of prerogative or right' over meat inspection.[44] Many MOsH remained anxious about their position and were not prepared to relinquish any responsibility. They dismissed veterinarians' claims to expertise and staunchly maintained that tuberculous meat was a public health issue, defending what they saw as their area of expertise. Nor were the public moved by veterinarians' calls for inclusion. MOsH had constructed their professional status around claims to expertise based on their epidemiological and bacteriological knowledge and their role as defenders of public health. Within the public health domain, the public did not accept the rival claims of veterinary surgeons who were struggling for professional status and remained associated with farriery. They did not recognise veterinary surgeons 'as being an authority on matters of meat inspection' because they had 'no certificate, that precious bit of paper, which on the face of it shows that he has had a training and passed examinations in those subjects'.[45] Clear boundaries existed between human and animal diseases, with the latter viewed

[40] GCA: Health committee, 29 Sept. 1896, E1/20/17.

[41] 'Inspection of Meat in Glasgow', *Sanitary Record*, 11 Apr. 1901, p. 319; GCA: Health committee, 20 June 1900, E1/20/21.

[42] See *Report of the Health of Liverpool during the Year 1893* (Liverpool, 1894), p. 45.

[43] 'Meat Inspection', *Public Health* xiv (1901/2), p. 727; 'The Swansea Meat Seizure Case', *Veterinary Record*, 23 Nov. 1901, p. 329.

[44] *Ibid.*, 29 Jul. 1905, p. 81.

[45] 'The Veterinary Profession and Public Health', *Veterinary Journal*, June 1908, p. 277.

essentially in agricultural and economic terms, which frustrated veterinarians' calls for greater involvement in public health.

Even when veterinarians were appointed, poor pay and their subordinate status to doctors engendered frustration. Veterinarians felt aggrieved that to legally inspect meat they had first to qualify as sanitary inspectors, a situation that reinforced their low status. In addition, meat inspection posts seldom paid more than £120 (and frequently less), and it was felt that 'no self-respecting qualified veterinarian [was] likely to apply for the post'.[46] MOsH tended to marginalise veterinary surgeons. They saw their veterinary colleagues as subordinate, often only calling them in when 'it is found necessary to go into Court'.[47] Veterinarians were seldom consulted over diseased meat, whilst their demands for greater authority and autonomy were frequently ignored. Conflicts between veterinary inspectors and sanitary inspectors, such as in Dumbarton, threatened the local system of meat inspection.[48] It was only with the 1913 Tuberculosis Order that the door seemed open to greater veterinary involvement in public health.

Inspecting meat

Although William Berry, MOH for Wigan, was clear that 'it is quite evident from the findings of the last Commission that we must retreat from the method of seizing all carcasses that are affected with tuberculosis', the first two commissions had a limited impact on the practicalities of meat inspection.[49] It took nearly a year after the second commission for the LGB to circulate guidelines for inspection. Rather than stipulating the nature of inspection, it tried to act as a facilitator. It would appear that not all areas of public health work by the end of the nineteenth century were characterised by compulsion. The LGB embraced the commissioners' recommendations, not least because it realised that a localised model was more acceptable on economic grounds owing to the large proportion of cattle which were felt to be tuberculous.[50] The guidelines outlined the official stance that inspectors should only condemn carcasses in which the disease was so generalised that no 'reasonable' amount of dressing could remove the tubercles. When the tubercles were confined to the organs only the diseased parts were to be removed and seized. This, it was felt, was sufficient to protect the public and at the same time evaded the difficult question of whether the bacilli were disseminated through the

[46] Vacher, *Handbook*, p. 3.
[47] GCA: Joint committee on inspection of dead meat, 30 Jan. 1891, MP29.171.
[48] *Sanitary Record*, 30 May 1913, p. 576.
[49] William Berry, 'Tuberculous Meat', *Public Health* xii (1899/1900), p. 284.
[50] *JRAS* ix (1898), p. 126.

glands and bloodstream that divided doctors and veterinarians.[51] The 1899 memorandum was followed by a further circular two years later in an attempt to offset concerns generated by Koch's statement to the 1901 British congress on tuberculosis. The LGB, alarmed by the continuing variations in inspection practices, used the opportunity to reaffirm its support for the findings of the second commission, adamant that the measures it proposed were adequate. However, whereas on the surface the guidelines issued by the LGB outlined a coherent pattern for inspection, much like early sanitary legislation, they lacked precision. Despite the LGB's growing commitment to food safety, its classification was tentative, out of step with practices in Europe where a more extensive system of inspection was in operation.

The guidelines issued by the LGB were welcomed by butchers and meat salesmen. In endorsing a localised model of infection they answered some of the grievances expressed by the meat trade and suggested a more moderate approach that would reduce the number of prosecutions. Sanitary inspectors and MOsH were less enthusiastic. Although they felt that the second commission had offered 'a fairly satisfactory basis [for] action', and were willing to use its findings to secure convictions, many opposed the LGB's guidelines.[52] They ran counter to the practices adopted by most MOsH who, it was felt, should be able to 'exercise [their] own judgement'.[53] The literature aimed at doctors contradicted the LGB's recommendations and reflected a generalised model that aided MOsH in their campaigns to clean up the meat trade. Although studies had revealed that tuberculous lesions were rarely present in the muscular system or flesh, opposition was vocal because the recommendations appeared to threaten a strict policy that allowed MOsH to condemn all tuberculous meat. *Public Health* repeated a widely held view among MOsH that whilst the risk of infection from sound meat from tuberculous animals was small, the entire carcass should be condemned. This was to prevent the public from being 'unwittingly' exposed to the risk of infection because 'it is impossible to demonstrate the absence of the bacillus and its products from any part of a tuberculous animal'. *Public Health* reinforced calls for total condemnation, arguing that it 'will do much to hasten the stamping out of tuberculosis among cattle'. The same stance was adopted at meetings.[54] It was felt that this offered a far better basis on which to protect the public and reform the meat trade.

Practices therefore continued to vary. Questions about whether tuberculosis was a localised or generalised disease persisted and encouraged a sense that it was 'impossible to lay down any regulations for the partial or total

[51] *Twenty-ninth Annual Report of the Local Government Board, 1899–1900* (London, 1900), p. 12. These recommendations have continued to shape responses to tuberculous carcasses.
[52] *Public Health* xi (1898/9), p. 455.
[53] Wellcome: SMOH council minutes, 21 Sept. 1900, SA/SMO/B.1/9.
[54] 'Condemnation of Tuberculous Meat', *Public Health* xii (1899/1900), pp. 706, 751.

condemnation of carcases on account of tuberculosis'.[55] Localism remained strong in public health work and it was left to each MOH 'to exercise his own opinion'.[56] Inspectors were not compelled to implement the LGB guidelines, and each inspector set his own standards in what was a highly subjective process, often dependent more on prejudice than any objective or scientific criteria. For example, the London boroughs were less inclined to accept the LGB's recommendations and continued to condemn all carcasses affected with tuberculosis. Edinburgh and Newcastle were equally stringent; Greenwich and Scarborough far less so, often not seizing carcasses even when they were emaciated.[57] It was these considerable variations, which the royal commissions had hoped to deal with, that ensured that meat inspection continued to be attacked as chaotic.

Despite local sensitivity to the interests of the meat trade, and a reluctance to prosecute when voluntary surrender could be secured, the renewed efforts to condemn tuberculous meat encouraged by the royal commissions prompted complaints. Decisions over meat became the most frequent source of contention between butchers, traders and inspectors, with 'members of the Fleshing Trade' only too ready to air their grievances – real or imagined.[58] With a growing number of prosecutions, the National Federation of Butchers and Meat Traders' Associations felt compelled to complain about the 'harsh proceedings of Medical Officers of Health who . . . confiscate in doubtful cases of tuberculosis animals'. Evidence was presented of butchers being unfairly prosecuted. Sanitary authorities were condemned for publicising cases before they had been heard. Complaints were voiced about harassment and 'ill-considered charges'. For the Federation, such practices were common. In response, demands were repeatedly made for protection from prosecution for those who had bought diseased meat in good faith and for compensation.[59] Individual butchers responded by abandoning the trade in English beef in favour of imported livestock. Others became more willing to conceal doubtful carcasses in the hope of avoiding prosecution.[60]

The level of anxiety expressed by butchers about prosecutions, 'unjust' or otherwise, did not mean that the system of meat inspection worked well. If Birmingham, Liverpool and Salford had efficient meat inspection, overall the

[55] *Select Committee on the Tuberculosis (Animals) Compensation Bill* (London, 1904), p. iii.
[56] 'Meat Branding', *Journal of Meat and Milk Hygiene* i (1911), p. 440.
[57] 'Seizure of Tuberculous Carcases', *Sanitary Record*, 12 Jan. 1911, p. 41; 'St Pancras Meat Case', *Public Health* xii (1899/1900), p. 710.
[58] GCA: Health committee, 18 Nov. 1901, E1/20/23; *Annual Report of the Medical Officer of Health on the Sanitary Condition of Warrington* (Warrington, 1909).
[59] 'The Brighouse Meat Case', *MTJCSG*, 17 Feb. 1898, p. 872; GCA: Health committee, 26 Aug. 1901, E1/20/22; CLRO: Report on importation and inspection of foreign meat, June 1905, PH/140.
[60] 'Tuberculous Meat', *Sanitary Record*, 7 Jan. 1904, p. 4; *Tuberculosis (Animals) Compensation Bill*, pp. iv, 103.

system was considered poor despite the ambitions of the royal commissions and the subsequent attempts to improve inspection.[61] The severity with which meat inspection was conducted depended on the district. Some areas were reluctant to act, as in Aberystwyth where the town council had little conception of its public health duties and preferred its inspectors not to do 'too much'.[62] The competence of sanitary and meat inspectors also varied greatly and this had a marked impact on inspection. Whilst some were considered 'capable and intelligent', painstakingly conducting their duties, a survey of local sanitary administration by the LGB between 1893 and 1894 revealed a picture of overwork, inadequate staffing and low pay.[63] Sanitary officials would neglect their meat inspection duties because of the burden of other work. As Francis Vacher noted in 1913, 'there is good reason to believe that very inadequate attention is given to food inspection, and that much food which is allowed to be exposed for sale is diseased'.[64]

Problems that had dogged meat inspectors in the late nineteenth century continued to frustrate action. Opposition from the meat trade and accusations against inspectors made inspection difficult. The number of inspectors was inadequate and opportunities to avoid detection great. Often the task was too large to make systematic examination possible and inspectors were forced to rely on voluntary surrender, tip-offs or assistance from the police. The *Journal of Meat and Milk Hygiene* was aware that it was not

> possible for him [inspectors] to devote the time necessary to efficient inspection, when we consider that most of the slaughtering in private slaughterhouses takes place at any time convenient to the butcher, which is usually before or after shop hours, when it would be inhuman to expect the inspector to attend after his arduous duties of the day.[65]

Butchers had also become adept at avoiding inspection. 'Stripping', 'rubbing' or otherwise tampering with carcasses remained common, although such practices also encouraged inspectors to be more suspicious. Too much inspection was left to chance: less scrupulous butchers strove to minimise their chances of detection by slaughtering and dressing diseased animals in rural districts before sending the meat to urban markets, or would 'slaughter wasted animals at night time when there is no risk of their operations being

[61] Gerald Leighton and Loudon Douglas, *The Meat Industry and Meat Inspection* vol. vi (London, 1905), pp. 1097, 1128, 1138.
[62] TNA: PRO, 'An Inquiry as to the Arrangements for the Discharge of the Duties of Inspectors of Nuisances', Sept. 1909, MH 96/405.
[63] *Report of the Inland Sanitary Survey, 1893–95, submitted by the Medical Officer of the Local Government Board* (London, 1896).
[64] Vacher, *Handbook*, p. 1.
[65] 'Meat Branding', *Journal of Meat and Milk Hygiene* i (1911), p. 440.

supervised'.[66] Some urban sanitary authorities did try to get round this problem by inspecting all meat entering their town or city, as in Glasgow.[67] However, the fault was felt to lie with rural sanitary districts.

Although some rural districts did attempt to police the meat trade, the system of rural meat inspection was considered 'a mere pretence'.[68] Considerable problems were seen to exist with the system of rural public health administration; deficiencies in meat inspection were one aspect of wider concerns. The quality of appointments in rural sanitary districts was widely considered to be low. Urban MOsH complained that no matter how hard they worked, the part-time and limited nature of inspection in rural areas, which was subordinate to private practice, meant that meat was being sent into towns after it had been 'tampered with, and stripped, and dealt with in every possible shape and form, so as to hide the disease'.[69] The size of rural districts made inspection difficult. The fact that rural slaughterhouses were scattered and often worked for only a few hours each week compounded the problem. Busy and under-staffed rural sanitary inspectors – some rural districts were only permitted to appoint one inspector – found it hard to fit meat inspection in with their other duties.[70] All too often meat escaped detection and made its way to nearby towns. Rural districts were believed to be conspicuously at fault, having in many cases wilfully neglected to apply sanitary legislation. Rural district councillors were accused of avoiding measures 'likely to affect [their] pockets' or offend farming interests.[71] With rural districts often dominated by conservative farmers, there was little chance that the situation could be improved.

It was not just in rural districts that problems occurred, however. Although the publicity surrounding bovine tuberculosis did encourage the voluntary surrender of tuberculous meat, MOsH were often forced to rely on moral coercion and it remained difficult to secure convictions. Differences between veterinarians and doctors, and between doctors, called to give evidence over the extent of infection that should be considered dangerous tended to confuse magistrates, frustrating efforts at conviction. As the *BMJ* explained, 'medical officers of health have cut from the dressed carcass portions which they have

[66] TNA: PRO, Memorandum, 24 Feb. 1914, MAF 101/358.

[67] J. Skinner, 'Difficulties of Sanitary Inspectors', *Journal of Sanitary Inspectors* i (1895–6), p. 332; see GRO: Cardiff property and markets committee, 21 Mar. 1906, BC/C/6/32.

[68] *Annual Report of the Medical Officer of Health for Caernarvonshire combined Sanitary Districts* (1911).

[69] *Royal Commission . . . into . . . Controlling the Danger to Man through the use as Food of the Meat and Milk of Tuberculosis Animals*, minutes of evidence, p. 99.

[70] *Annual Report of the Medical Officer of Health for Roxburgh Rural District Council*, (1911); TNA: PRO, Report to the LGB upon the sanitary circumstances and administration of the Brecknock Rural District, 1907, MH 96/605.

[71] *Report and Special Report from the Select Committee on Housing of the Working Classes Acts Amendment Bill*, PP (1906) ix; A. Mearns Fraser, 'Diseased Meat and Milk', *Nineteenth Century and After* lxii (1907), p. 243.

demonstrated to be swarming with bacilli, and still magistrates continue of the opinion that the flesh is not affected'.[72] All too often, 'the interests of public health' were 'being very seriously subordinated to those of the meat trade'.[73]

Into *The Jungle*

If domestic meat presented one set of problems, at the start of the twentieth century imported meat raised other issues. Foreign meat had traditionally been considered to present a lower risk of tuberculosis 'on account' of the cattle leading a healthy, open-air life.[74] Such claims removed the necessity to inspect imported meat, which, because of its refrigerated state, was problematic. However, the introduction of import controls across Europe in the 1890s to combat the spread of bovine tuberculosis did put pressure on Britain to act as part of an international effort against the disease. A rise in the amount of imported meat between 1896 and 1905 made inspection of foreign meat essential, especially as it was increasingly felt that 'much imported meat is actually diseased'. However, it was the publication of Upton Sinclair's novel *The Jungle* in 1906, and the resulting food scare over meat from North America, which forced action.[75]

Imports of North American beef to Britain had reached a peak in 1900 and had become associated with high quality, but *The Jungle* revealed a very different picture by highlighting the appalling conditions in the meat packing plants of Chicago. Systematic meat inspection in Chicago had only been adopted in 1904, and in the first year nearly 4 million pounds of meat was seized as unfit, 80 per cent of which 'on account of tuberculosis'.[76] In the United States, Sinclair's account produced popular indignation and revulsion. In Britain, concern had already been expressed in *The Lancet* about the meat packing and slaughtering trade in Chicago the year before, but it was the publication of *The Jungle* that concentrated attention on the potential dangers of the foreign meat trade.[77] Despite attempts in the United States to clean up slaughterhouses and tighten inspection, British critics felt that these efforts were cosmetic and that the Federal inspection of meat was little better than

[72] 'The Meat of Tuberculous Cattle', *BMJ* i (1897), p. 307.

[73] 'Condemnation of Tuberculous Meat', *Public Health* xii (1899/1900), p. 706.

[74] *Tuberculosis (Animals) Compensation Bill*, p. 99.

[75] George Newman, 'The Administrative Control of Food', *Public Health* xix (1906), p. 85; Herbert Maxwell, 'Tuberculosis in Man and Beast', *Nineteenth Century* xliv (1898), p. 679.

[76] 'Inspection of Meat in Chicago', *Veterinary Record*, 7 July 1906, p. 5; Jon A. Yoder, *Upton Sinclair* (New York, 1975); James Harvey Young, *Pure Food: Securing the Federal Food and Drugs Act of 1906* (Princeton, 1989), pp. 233–51.

[77] *Lancet* i (1905), pp. 49–52, 120–3, 183–5, 258–60.

a sham, as vested interests continued to be protected.[78] As consumer confidence in American meat fell, protectionist sentiments were voiced as part of debates about tariff reform. At the same time, the nature of British beef was defended. The example of Chicago was used to argue for improvements in Britain and helped 'arouse the public to a sense of the dangers which they run by eating diseased and unwholesome meat'.[79]

George Seaton Buchanan at the LGB, head of the newly formed Foods Section, exploited the scandal. The Foods Section had been established in 1905 and had already produced reports on methods of reducing the risk of importing diseased meat. The LGB had also tried to acquire unrestricted regulatory powers over food purity and had launched further investigations into food-borne infections. Buchanan consulted MOsH responsible for food inspection in the Port of London, who repeated long-standing grievances that existing measures were inadequate.[80] Buchanan used the evidence to call for new centralised regulations, which were incorporated into the 1907 Public Health (Regulation as to Food) Act. The Act allowed the LGB to make public health regulations for the importation, preparation, storage and distribution of food and drink. As such, it 'offered the prospect of an unprecedented degree of central government intervention in the regulation of food provision'. Although the Act was resisted because of the perceived anti-democratic provisions that allowed the LGB to intervene where local authorities failed to inspect meat, it was pushed through to enable port authorities to deal with diseased meat.[81]

Measures had been introduced following the cattle plague in the 1860s to prevent the importation of diseased livestock from 'scheduled countries' – those where particular diseases were known to be endemic, or where a recent outbreak had occurred. This had ensured that all imported animals were slaughtered at the port of landing, although there was no obligation to inspect the carcasses.[82] In addition, ports were covered by sanitary legislation that related to the removal of nuisances and hence contained provision for meat inspection. Under the 1872 Public Health Act, separate Port Sanitary Authorities were created, whilst the 1896 Foreign Animals Order required any suspicious cases to be referred to the sanitary authority. Although these measures had support from the meat trade, controls were incomplete.[83] Local

[78] Alfred Leffingwell, *American Meat, and its Influence upon the Public Health* (London, 1910), p. 8.

[79] 'Meat Inspection Question in Bermondsey', *Veterinary Record*, 27 Oct. 1906, p. 265.

[80] TNA: PRO, Note of discussion with London MOsH, 11 June 1906, MH 113/48.

[81] Michael French and Jim Phillips, *Cheated Not Poisoned? Food Regulation in the United Kingdom, 1875–1938* (Manchester, 2000), pp. 87–92.

[82] *Annual Report of the Veterinary Department of the Privy Council Office for the Year 1878* (London, 1879), p. 3.

[83] *Departmental Committee Appointed to Inquire into Pleuro-pneumonia and Tuberculosis in the United Kingdom*, PP (1888) xxxii, p. xxv.

authorities, already overburdened by other sanitary work, devoted little attention to inspecting imported cattle. In practice, only those diseases covered by the Contagious Diseases (Animals) Act attracted the attention of inspectors at foreign meat markets.[84] The 1907 Act and the 1908 Foreign Meat Regulations that followed were designed to prevent this by extending the powers available to local and port sanitary authorities to inspect and destroy diseased meat. Customs officers were required to notify sanitary officials if any ship was found to have a cargo that 'comprises foreign meat'.[85]

However, the regulations did not find approval, since customs officials lacked the expert knowledge to identify diseased meat. Inspectors were frequently only called in to certify food as not fit for consumption rather than to inspect meat. Un-classed meat was particularly problematic, as 'in cases of meat cut up into small pieces no guarantee can be obtained that it has not come from animals the subject of disease in some form or other'.[86] However, the regulations did have a marked impact on the amount of foreign meat seized. Progressive Port Sanitary Authorities started to appoint separate meat inspectors to examine all imported meat.[87] Shipping companies began to complain about the stringency with which the regulations were enforced. They objected to frozen meat being defrosted and inspected. Reports suggested that imported meat sold in Britain was, because of the regulations, far less likely to be tuberculous, although at a local level inspectors reported 'revolting revelations' when meat was defrosted.[88]

Events in Chicago renewed interest in the nature of meat inspection in Britain. Many concluded that considerable improvements were still needed. As the editor of the *Veterinary Journal* commented, 'We have no need to go to Chicago to look for remediable faults in our food supply.'[89] It was noted with alarm that MOsH were frequently too overworked to concern themselves with meat inspection, whilst veterinarians reiterated longstanding complaints that doctors were unable to properly inspect live animals or 'recognise many of the diseases with which animals may be affected'.[90] Interest groups including the National Health Society campaigned for improvements to food inspection and pressure was applied on the LGB to extend local authorities' powers over the meat trade.[91] However, little concerted action was taken beyond the regulation of imported meat and efforts by sanitary authorities to improve inspection.

[84] *Royal Commission . . . into . . . Controlling the Danger to Man through the use as Food of the Meat and Milk of Tuberculosis Animals*, minutes of evidence, p. 72.
[85] Leighton and Douglas, *Meat Industry*, p. 1584.
[86] CLRO: Port of London sanitary committee report (1906), p. 59.
[87] See GRO: Cardiff health and port sanitary committee, 5 Jan. 1909, BC/C/6/37.
[88] 'Tuberculous Frozen Meat', *Medical Officer*, 12 Apr. 1913, p. 165; GCA: Report of the veterinary surgeon, 1908–9, C2/1/9.
[89] 'Veterinary Inspection of Our Food Supply', *Veterinary Journal*, Sept. 1906, p. 473.
[90] H. Woodruff, 'Meat and Milk Inspection', *Public Health* xix (1906/7), p. 99.
[91] 'The Meat and Milk Supply', *Sanitary Record*, 1 Nov. 1906, p. 409.

Despite the ambitions of the second royal commission, meat inspection remained chaotic, beset with the same problems that had worried sanitarians since the 1870s. Identification and removal remained the cornerstones of public health policy on tuberculous meat, but without compensation, uniform standards, and moves to limit the disease in cattle, it remained only a partial solution. With every inspector a law unto himself, practices continued to vary between sanitary authorities. Meat inspection was routinely attacked for its perfunctory nature, while sanitary officials saw it as onerous and fraught with difficulties. 'Now and again', explained the MOH for Stepney, 'an offender is caught and fined heavily, but the proportion so detected is infinitesimal compared with those who escape'.[92] The only solution seemed to be better inspection through the provision of public slaughterhouses.

Public slaughterhouses

Slaughterhouses were part of the urban fabric, and by the nineteenth century the visceral horrors of the slaughterhouse had come to represent a distasteful, dirty environment that offended sensibilities and posed numerous public health problems. In some ways, they represented an archetypal nuisance. Few slaughterhouses were purpose-built: many were found in 'ordinary limited back premises of a small third-rate house, sometimes of a wash-house or back kitchen; at other times of a roofed in back yard'.[93] In these 'dismal dungeons', cellars and back alleys, dirty and diseased conditions were widely reported, with contemporaries pointing to an atmosphere in which disease could readily spread. Slaughterhouses were regularly described as 'objectionable' where the stench from the 'faecal matter, the guts, the blood and the skins of animals' was 'intolerable'.[94] For public health reformers and middle-class commentators, private slaughterhouses represented an urban nuisance that had to be cleansed and a source of cruelty that needed to be repressed.

The need for reform was pushed by medical and veterinary bodies and by voluntary organisations campaigning for the suppression of private slaughterhouses to promote humane slaughtering practices. By the 1820s, the treatment of animals had become a national preoccupation, but the emergence of tuberculosis as a zoonotic disease recast interest in slaughterhouses. Although their sanitary and humanitarian dimension remained important, debate increasingly focused on the relationship between slaughterhouses and meat inspection. A language of diseased meat control began to take over attempts

[92] 'London Food Horrors', *ibid.*, 28 June 1906, p. 556.

[93] *Public and Private Slaughterhouses: Report addressed to the Society for Providing Sanitary and Humane Methods for Killing Animals for Food* (London, 1882), p. 4.

[94] Thomas Dunhill, *Health of Towns: A Selection of Papers on Sanitary Reform* (London, 1848), p. 20; GCA: Complaints about the cleanliness of slaughterhouses, A2/1/3/12.

to regulate local slaughterhouses, reflecting a shift in rhetoric in public health work away from a doctrine of nuisances. As the *BMJ* explained in 1895, 'the greatest danger' arose not from insanitary slaughterhouses but 'from the failure to detect and destroy all diseased and unwholesome carcasses'. The journal worried that consequently 'much [diseased] flesh was constantly offered for sale'.[95] Private slaughterhouses were considered secretive places, offering 'extraordinary facilities' for the slaughtering of diseased animals and for the disposal of unwholesome meat, where anything could be done without fear of detection.[96] The sheer number of private premises ensured 'the impossibility of that complete inspection which is necessary' for the detection of diseased meat, especially as they were often widely distributed across a town.[97] In addition, most had insufficient lighting or room for an adequate examination of carcasses. Diseased organs could be concealed or removed before the inspector arrived.[98] Under these conditions, private slaughterhouses became vital components in the traffic of diseased meat.

Public abattoirs offered a solution to these problems, and sanitarians and veterinarians turned to Europe for inspiration. Public slaughterhouses had been established in Prussia from the late 1870s onwards, and by the 1890s most German towns had a municipal abattoir.[99] Public health journals referred to the German system as 'perfect' and hoped that 'the movement for establishing public abattoirs will greatly and speedily spread in our country'.[100] It was argued that public abattoirs not only encouraged greater cleanliness and more humane methods of dispatching animals, but also allowed for the efficient inspection of all animals before and after slaughter. It was felt that this would discourage butchers from succumbing to the temptation to mask the signs of disease, and promote the constant vigilance seen as essential in combating bovine tuberculosis to ensure that 'the most valuable indications whereby the nature of the meat may be judged' would not be removed.[101] Public slaughterhouses were felt to offer 'the only means of obviating the risk of animal diseases being communicated to man from the consumption of unsound meat'.[102] It was

[95] 'The Need for Public Slaughterhouses', *BMJ* i (1895), p. 1015.

[96] *BMJ* i (1873), p. 414.

[97] CLRO: Sanitary committee, 2 June 1874.

[98] William Howarth, *Meat Inspection Problems, with special reference to the Development of Recent Years* (London, 1918), p. 32.

[99] Marjatta Hietala, 'Hygiene and the Control of Food in Finnish Towns at the Turn of the Century: A Case Study from Helsinki', in *The Origins and Development of Food Policies in Europe*, ed. John Burnett and Derek J. Oddy (Leicester, 1994), p. 124.

[100] 'Food Inspection', *Sanitary Record*, 26 July 1895, p. 81.

[101] T. Morrison Legge and Harold Sessions, *Cattle Tuberculosis: A Practical Guide to the Farmer, Butcher, and Meat Inspector* (London, 1898), p. 45.

[102] T. Morrison Legge, 'Some Points of Difference between English and Continental Methods of Municipal Sanitary Administration', *Public Health* vii (1894/5), p. 123; 'The

anticipated that with them 'the traffic in diseased meat would soon sink within insignificant limits'.[103]

Although there was growing pressure for public abattoirs from doctors and veterinarians, progress was uneven; the results mixed. Early local efforts to establish corporation markets and slaughterhouses, as in Cardiff in 1835, were initially designed to combat the perceived urban nuisances posed by private slaughterhouses and did not necessarily mean that inspection occurred or was rigorous.[104] Although the 1875 Public Health Act granted sanitary authorities the ability to open public abattoirs, it was the Glasgow trial that stimulated debate on the need for public slaughterhouses. Some local authorities began to move in this direction. For example, a public slaughterhouse was opened in the Dyfatty area of Swansea in 1889 with the specific aim of preventing the sale of diseased meat. Provision was further encouraged by the second royal commission on tuberculosis, which considered the creation of public abattoirs 'a necessary preliminary to a uniform and equitable system of meat inspection'.[105] In supporting municipal abattoirs, the commissioners voiced a growing consensus that they were vital to the effective inspection of meat. Concerns about unsanitary conditions now took second place to the advantages of public slaughterhouses to meat inspection.[106] Debates about meat inspection and municipal abattoirs combined with anti-vivisection interests and organisations campaigning for the better treatment of animals to press for public slaughterhouses. Influenced by these debates, a growing number of sanitary authorities started to act, with all the large towns in Scotland replacing private slaughterhouses with public abattoirs by 1897.[107] Provision remained patchy in England, but by 1911 'scarcely a week goes by but some local authority or other discusses the question of the erection of a public abattoir'.[108]

However, despite the interest shown, 'time after time' schemes for public abattoirs were 'enthusiastically approved' and 'never realised'.[109] The LGB felt that local councillors were apathetic because of the strength of the meat lobby and cost of provision, whilst local authorities were frustrated by the existing registration of private slaughterhouses.[110] Although the sanitary and health benefits of public abattoirs were accepted, their provision remained

Abolition of Private Slaughterhouses', *JCPT* ii (1889), p. 343; 'Slaughterhouse Reform', *Sanitary Record*, 15 Aug. 1884, p. 64.

[103] 'Diseased Meat', MTG, 5 Apr. 1862, p. 352.

[104] GRO: Cardiff local board of health, 26 July 1861; 27 June 1862, BC/LB/3.

[105] *Royal Commission . . . into . . . Controlling the Danger to Man through the use as Food of the Meat and Milk of Tuberculosis Animals*, part i, p. 10.

[106] See *Annual Report of the Medical Officer of Health for Cheltenham* (1899).

[107] *Royal Commission . . . into . . . Controlling the Danger to Man through the use as Food of the Meat and Milk of Tuberculosis Animals*, minutes of evidence, p. 147.

[108] 'Public Abattoirs', *Journal of Meat and Milk Hygiene* i (1911), p. 155.

[109] 'Organised Opposition to Public Slaughter-Houses', *Medical Officer*, 6 July 1912, p. 1.

[110] TNA: PRO, Paddison to Jerrod, 24 Apr. 1914, MAF 101/358.

controversial. Often the attack was led by local butchers. In Manchester, the local Butchers' Guardian Association resisted the opening of a public abattoir, despite the obvious public health dangers of the existing private slaughterhouses from which blood and offal 'drained unchecked into the sewers'.[111] In Middlesbrough, efforts to abolish private slaughterhouses by the chief sanitary inspector induced the Butchers' Association to secure the election of two councillors to oppose him.[112] Opposition was more than just a local phenomenon, however. The National Federation of Butchers and Meat Traders' Associations claimed that butchers would have to be evicted from private slaughterhouses and then forced to transfer their business to public abattoirs, as few would willingly use them.[113] Complaints were made that municipal abattoirs were expensive; that any move to close private slaughterhouses would drive up the price of meat as butchers would be forced to move from premises that cost them little and were near their shops.[114] Butchers also bemoaned that a public abattoir 'destroys the privacy of a man's business' and fostered gossip which damaged business.[115] Butchers, as Perren has noted, appealed to a sentiment that 'represented the small scale trader as being locked in a struggle for his very existence with forces threatening to exert a powerful monopoly'.[116] Yet, despite these claims, at the root of their opposition was a desire to avoid systematic inspection. Local authorities were therefore wary of pushing ahead with public abattoirs in the face of such opposition.[117]

Fears of increased vigilance and control were real. In those towns in which public slaughterhouses were opened, there was marked improvement in meat inspection. The opening of the public abattoir in Bury in 1902, and moves in the following year to close all private slaughterhouses, for example, led to a dramatic rise in the number of animals inspected and amount of diseased meat destroyed.[118] However, problems remained. Municipal abattoirs could only have a marked impact on meat inspection when all private slaughterhouses in a city or borough were closed, a step few local authorities were prepared to take. Butchers, in order to escape inspection, erected private slaughterhouses outside city boundaries. They resisted moving their businesses, preferring their own premises 'however small and inconvenient' where inspection was far less

[111] A. Redford, *A History of Local Government in Manchester* (London, 1939), p. 289.
[112] *Sanitary Inspectors' Journal* v (1899/1900), p. 164.
[113] *Royal Commission . . . into . . . Controlling the Danger to Man through the use as Food of the Meat and Milk of Tuberculosis Animals*, minutes of evidence, p. 33.
[114] LMA: LCC public health committee, 3 Nov. 1898.
[115] J. Garrett, 'On Public and Private Slaughterhouses', *Public Health* vii (1894/5), pp. 387.
[116] Richard Perren, *The Meat Trade in Britain, 1840–1914* (London, 1978), p. 155.
[117] For example, the LCC repeatedly delayed opening public slaughterhouses 'owing to the clamour made by the meat trade and the fear of losing trade votes': 'The LCC and Private Slaughterhouses', *Sanitary Record*, 6 Aug. 1903, p. 161.
[118] A. Brindley, 'Public Abattoirs', *Public Health* xviii (1905/6), p. 444.

organised.[119] Others would simply transfer their businesses to less well-regulated districts. One writer on meat inspection pessimistically noted in 1918 that given these problems, 'there seems no possibility in the near future of public slaughter-houses replacing private ones'.[120]

Diseased meat control

By the last decade of the nineteenth century commentators began to suggest that the campaigns against tuberculous meat were forcing a growing number of butchers and cattle dealers 'to go into foreign beef to the detriment of the home producer' and the consumer.[121] However, despite attempts to improve inspection and establish public abattoirs, the widespread systematic sale of diseased meat was still considered commonplace. Contemporaries worried that the consumer was 'daily eating the flesh of diseased animals, and is unaware of it'. Controls were considered inadequate; in some localities non-existent. As W. Barnes, the veterinary inspector and superintendent of abattoirs in Islington, noted 'There can be no gainsaying the fact that the system of meat inspection in this country is most unsatisfactory, and must continue so until those authorities responsible for public health ignore trade objects and vested interests.'[122] With the system of inspection unable to prevent the sale of diseased meat, attention increasingly came to focus on eradicating bovine tuberculosis in cattle.

[119] A. Jasper Anderson, 'Public Slaughterhouses and Meat Inspection', *Public Health* vii (1894/5), p. 381.
[120] Howarth, *Meat Inspection*, p. 34.
[121] *Royal Commission . . . into . . . Controlling the Danger to Man through the use as Food of the Meat and Milk of Tuberculosis Animals*, minutes of evidence, p. 175.
[122] Cited in *Bibby's Book on Milk*, p. 128.

9

Death in the milk pail

Tuberculosis and milk

Historians have considered milk the prime factor in shaping fears about bovine tuberculosis. It has assumed an important position in writing on the infant and child welfare movement, and in research on adulteration, with recent scholarship linking milk to the emergence of a new politics of consumption in the twentieth century.[1] In examining the problem of milk and tuberculosis, attention has concentrated on the difficult adoption of pasteurisation.[2] Certainly, alarm about bovine tuberculosis did drive efforts to control the milk trade, building upon anxieties that became apparent in the 1850s about milk hygiene and the role it played in the transmission of disease. At first comparisons were made with water, but the emergence of bovine tuberculosis as a zoonosis encouraged a shift in interest. Milk no longer just spread disease: it could be intrinsically dangerous. By the late 1890s, bovine tuberculosis had come to dominate discussions about milk and vice versa at an international level.[3] In Britain, the fears aroused by tuberculous milk merged with anxieties about degeneration and child health and the vulnerability of children to infected milk became a major issue. The risk, however, was not just to human health: bovine tuberculosis became 'the most serious menace to the dairy industry'.[4] By 1900, the dairy industry had emerged as the cornerstone of Britain's agricultural industry and the trade in country milk to towns had

[1] See Peter J. Atkins, 'White Poison? The Social Consequences of Milk Consumption, 1850–1930', *SHM* v (1992), pp. 207–27; Deborah Dwork, 'The Milk Option: An Aspect of the History of the Infant Welfare Movement in England, 1898–1908', *Medical History* xxxi (1987), pp. 51–69; Jim Phillips and Michael French, 'State Regulation and the Hazards of Milk, 1900–1939', *SHM* xii (1999), pp. 371–88; Frank Trentmann, 'Bread, Milk, and Democracy in Modern Britain: Consumption and Citizenship in Twentieth-Century Britain', in *The Politics of Consumption*, ed. Martin Daunton and Matthew Hilton (Oxford, 2001), pp. 129–63.

[2] Peter J. Atkins, 'The Pasteurisation of England: The Science, Culture and Health Implications of Milk Processing, 1900–50', in *Food, Science, Policy and Regulation in the Twentieth Century: International and Comparative Perspectives*, ed. David F. Smith and Jim Phillips (London, 2000), pp. 37–51.

[3] See Katherine McCuaig, *The Weariness, the Fever, and the Fret: The Campaign against Tuberculosis in Canada, 1900–1950* (Montreal, 1999); Barbara G. Rosenkrantz, 'The Trouble with Bovine Tuberculosis', *BHM* lix (1985), pp. 155–75.

[4] H. W. Conn, *Practical Dairy Bacteriology* (London, 1918), p. 88.

attained 'enormous dimensions'.[5] Outbreaks of the disease and attempts to impose controls on the milk trade appeared to threaten the financial health of the dairy industry, encouraging calls from the strong farming lobby for protection. The result was a potent mixture of interests that made milk from tubercular cows a national issue and a focus for reform.

Part of the reason for a shift in interest from meat to milk came from a sense that by 1900 the problem of diseased meat was on the way to being solved. Confidence was expressed that the abolition of private slaughterhouses, the establishment of public abattoirs, and efficient meat inspection were sufficient to prevent the sale of tuberculous meat 'because it would no longer be remunerative to keep tuberculous cows until they become seriously diseased'.[6] Studies demonstrating that cooking rendered diseased meat safe provided further reassurance. However, milk was less amenable to regulation. Urban sanitary authorities increasingly worked to prevent the sale of milk from tuberculous cows, but controls on the conditions under which dairy cows were housed, the inspection of cattle, and measures to purify milk were difficult to enforce. Tensions between sanitary officials, veterinarians, farmers and the state served to frustrate attempts to secure a safe milk supply. The milk question was not amenable to a simple answer.

'Death in the milk pail'

Rising milk consumption from the 1850s onwards, falling prices and the growing importance of tea and coffee drinking saw milk become an important article of diet.[7] However, the reputation of the milk trade remained low throughout the nineteenth century. Milk was widely believed to be the most commonly adulterated food, and was often mixed with water or preservatives. Concern went beyond this as milk was gradually identified as a 'suitable medium for the growth of micro-organisms'.[8] Even before laboratory studies determined the bacterial content of milk, it had been clear to many sanitarians from how milk was produced that it could cause or transmit disease. Even with limited methods of analysing milk, studies revealed that it frequently

[5] David Taylor, 'The English Dairy Industry, 1860–1930: The Need for Reassessment', *AHR* xxii (1974), p. 153; J. Prince Sheldon, *Live Stock in Health and Disease* (London, 1902), p. 332.

[6] Arthur Newsholme, *Hygiene: A Manual of Personal and Public Health* (London, 1906), p. 24.

[7] Derek J. Oddy, 'A Nutritional Analysis of Historical Evidence: The Working Class Diet, 1880–1914', in *Making of the British Diet*, ed. Derek J. Oddy and Derek Miller (London, 1976), p. 221.

[8] W. Howarth, 'Influence of Insanitary Milk Supplies on a Town's Health', *Sanitary Record*, 22 July 1898, p. 81.

contained high levels of impurities and bacteria.[9] The way milk was produced and sold created what Trentmann has referred to as the 'long chain from cow to consumer' that made milk 'a haven for bacilli'.[10] Milk could be polluted with dirt, infectious agents or other substances at the farm, during transit, where it was sold, and in the home. Milking practices were compared to a 'process of unscientific inoculation' with germs, and dairies were frequently described as dirty and unhygienic.[11] Under these conditions, and because of the strong link made between dirt and disease, contemporaries readily assumed that milk was an important source of contagion.

If by the 1890s bacteria and germs had come to represent a hidden enemy in the milk pail, investigations in the 1850s had already raised questions about the safety of milk as contemporaries speculated that milk from diseased cows sowed the seeds of disease.[12] This belief that milk might be responsible for the transmission of zymotic diseases gained ground throughout the 1860s and 1870s. Comparisons were made with water-borne disease based on epidemiological evidence that suggested that milk was a vehicle of contagion. Milk was quickly implicated in the transmission of a number of infectious diseases, notably typhoid, diphtheria and scarlet fever.[13] Although milk-borne outbreaks of these diseases were not the leading source of infection, apprehension was expressed about the 'constant danger' the public were exposed to through the 'introduction and propagation of infectious or contagious diseases through the medium of milk'.[14] By 1881, as *The Lancet* explained, 'instances of the spread of disease by infected milk are becoming numerous' with milk blamed for ninety-five outbreaks of epidemic disease between 1881 and 1896.[15] Local doctors became certain of 'the peculiar capacity of that article of diet for absorbing and propagating certain infectious diseases'.[16] Unhygienic conditions in milk production made it an easy target. In some cases, cows' milk was blamed for epidemics mainly because investigators were disposed to jump to conclusions, having eliminated other factors.[17] MOsH found it gratifying to identify

[9] See A. Houston, *Report on the Bacteriological Examination of Milk* (London, 1905).
[10] Trentmann, 'Bread, Milk, and Democracy', p. 139.
[11] Frederick Dodd, *The Problem of the Milk Supply* (London, 1904), p. 13.
[12] MTG, 11 Apr. 1857, pp. 364–5.
[13] Edward Ballard, *On a Localised Outbreak of Typhoid Fever in Islington . . . traced to the use of impure milk, etc* (London, 1871); *Report of the Medical Officer of the Local Government Board* (London, 1873), p. 193; Alfred Smee, *Milk, Typhoid Fever, and Sewerage* (London, 1873).
[14] Ernest Hart, 'The Influence of Milk in Spreading Zymotic Disease', *Transactions of the International Medical Congress* iv (London, 1881), pp. 491–544; W. Leslie MacKenzie, 'The Hygiene of Milk', *EMJ* v (1899), pp. 372–8; GCA: Health committee, 4 Nov. 1878, E1.20/3.
[15] 'Infected Milk', *Lancet* i (1881), p. 27; *The Municipalisation of Milk*, Fabian Tract No. 90 (London, 1899), p. 2.
[16] GRO: Cardiff health and port sanitary committee, 8 May 1888, BC/C/6/7.
[17] See *Annual Report of the Medical Officer of the Local Government Board* (London, 1886);

some source for epidemic outbreaks of disease, especially following work that implied that water was mostly free of pathogenic germs. The dairy industry interpreted the odium attached to milk as a 'convenient excuse for ignorance of a cause of infection'.[18] However, work on scarlet fever did push home the idea that in some cases, milk was not just the medium for contagion but the cow the source of infection. By identifying dairy cows as a potential source of disease, the idea that some animal diseases were zoonotic was reinforced.[19]

By the 1890s, the medical community was in the awkward position of accepting the idea that milk, under certain circumstances, was a threat to health because it offered 'a rich culture for the growth of micro-organisms' but at the same time had nutritional benefits, especially for nursing mothers and infants.[20] Milk-borne infectious diseases therefore attracted considerable publicity, aided by the fact that diseased milk was no respecter of market forces. With many infants and children given cows' milk and believed to be more susceptible to infection, the threat presented by diseased milk was incorporated in debates on physical deterioration and imperial decline that were gaining momentum in the 1890s and 1900s. Fears of degeneration provided a potent context and language within which to discuss existing concerns about disease and milk. As public anxiety about the prospect of national deterioration escalated following the Boer War, concern about milk as a potential factor in that deterioration increased. Dirty and infected milk was blamed for corrupting the health of British children at a time when improvements in child health were considered vital to the maintenance and enhancement of the nation. Some writers suggested that 50 per cent of the total infant mortality rate was due to disorders of the digestive system that could be directly linked to dirty or diseased milk.[21] Fears about milk and disease merged with alarm about the hygienic quality of milk as reports on the milk trade revealed widespread adulteration and insanitary practices. For the Interdepartmental Committee on Physical Deterioration, milk was dangerous to health and 'every effort should be made to get clean milk'.[22] It was a view reflected in the popular and medical press.

ibid. (London, 1887), p. xiii; GRO: Cardiff health and port sanitary committee, 11 Sept. 1888, BC/C/6/7.

[18] 'In Defence of Milk', *Cowkeeper and Dairyman's Journal*, Apr. 1904, p. 233.

[19] *Transactions of the Epidemiological Society* v (1885–6), p. 104; *Report of the Medical Officer of the Local Government Board* (London, 1883), p. 327.

[20] Albert E. Bell, *The Pasteurisation and Sterilisation of Milk* (London, 1899), p. 9.

[21] George Newman, *Infant Mortality: A Social Problem* (London, 1906); James Niven, *On the Improvement of the Milk Supply of Manchester* (London, 1896), p. 4.

[22] 'Physical Deterioration', *BMJ* i (1904), pp. 319–20.

Milk and the 'germs of consumption'

By 1900, diseased milk and tuberculous milk had become interchangeable in much the same way as diseased meat and tuberculous meat had become synonymous. Milk had been first identified as a potential vector in the transmission of bovine tuberculosis in 1846, and by the 1880s the disease had emerged as the main 'terror . . . [in] the milk can'.[23] Just as with experiments on the transmissibility of bovine tuberculosis from cows to humans, the lead in identifying meat as an agent of infection was taken by European pathologists as they amassed evidence that bovine tuberculosis could be spread from cows to humans.[24] In Britain, the idea started to be expressed in the 1850s that it is not improbable that 'from having sometime or another partaken of milk drawn from cows that were diseased' tuberculosis could result 'where there previously existed no hereditary taint'.[25] Evidence from studies of cows in Parisian dairies was used, but initially the idea that milk could cause tuberculosis remained marginal. It was only in the late 1870s that observational evidence began to verify that 'consumption and similar maladies may be transmitted to man by the ingestion of milk from diseased cows'.[26] Although there were dissenting voices that suggested that these findings were inconclusive, fears about milk were increasingly supported by bacteriological studies that revealed high levels of the bacilli in urban milk supplies. These highlighted 'how easily and unsuspectingly that article of human food may become a purveyor of wholesale mischief'.[27] If an 'epidemic of consumption' from milk was considered unlikely, the view that milk was an important source of infection for tuberculosis was endorsed by the first and second royal commissions on tuberculosis. The commissioners emphasised the threat to infants and children under 5 as part of growing concerns about child health.[28] Although

[23] *Lancet* i (1888), p. 638.
[24] *Annual Report of the Agricultural Department of the Privy Council Office, on the Contagious Diseases, Inspection, and Transit of Animals for the Year 1888* (London, 1889), p. 39; Charles Creighton, *Bovine Tuberculosis in Man: An Account of the Pathology of Suspected Cases* (London, 1881), p. 9; 'Can the Milk of Phthisical Cows Produce Tuberculosis?', *Veterinary Journal* vi (1878), pp. 188–97.
[25] H. Hodson Rugg, *Observations on London Milk, shewing its unhealthy character* (London, [1850]), p. 9.
[26] George Fleming, 'The Transmissibility of Tuberculosis', *British and Foreign Medico-Chirurgical Review* liv (1874), pp. 461–86; Alexander Wynter Blyth, *The Composition of Cows' Milk in Health and Disease* (London, 1879), p. 8.
[27] See George Bantock, *Modern Dictionary of Bacteriology* (London, 1899); *Tuberculous Meat: Proceedings at Trial under Petitions at the Instance of the Glasgow Local Authority against Hugh Couper and Charles Moore* (Glasgow, 1889), p. 213; *Journal of Agriculture* (1888), p. 223; MTG, 10 Nov. 1883, p. 550; 'Tuberculosis', *MTJCSG*, 27 Oct. 1888, p. 11.
[28] *Royal Commission Appointed to Inquire into the Administrative Proceedings for Controlling*

questions were asked in the press about 'How far may a cow be tuberculous before her milk become[s] dangerous as an article of food?', pointing to the ongoing confusion as to whether the disease was localised or generalised, the general view by the 1890s was that milk was the perfect vector for the transmission of the tubercle bacilli.[29]

Farmers and dairymen were harder to convince. They continued to 'scoff at the alleged danger of tuberculous milk' and denounced 'all efforts to avert such danger as the outcome of a mere craze'.[30] The agricultural press played down the risk. The *Dairyman's Record* explained that the hazard was 'not so great as has been generally supposed', claiming that 'at all times when the udder is not affected, the danger from the use of the milk is quite limited'.[31] The *Cowkeeper and Dairyman's Journal* therefore strongly advised the public not to 'believe all the doctors tell you' with regard to milk.[32] Such reactions earned farmers and dairymen a reputation for indifference to safeguarding the public.

A number of influential doctors did play down the risk from milk sceptical of the contagious properties of tuberculosis, but sceptics could do little to counter the consensus that milk was the 'vehicle by which tubercle bacilli are often introduced into the bodies of human beings', even if it was not the main means by which the disease was introduced.[33] Although it was felt that the common practice of mixing milk probably reduced the threat of infection, many MOsH could confidently claim by 1899 that because levels of bovine tuberculosis were alarmingly high in dairy cows most milk was 'loaded with the germ of tuberculosis' and had to be considered dangerous.[34] It was feared that such high levels of the bacilli resulted in 'an enormous stream of infectious milk . . . pouring [into] our cities'. Empirical evidence backed up this stance as local sanitary authorities conducted investigations into the milk supply: nationally it was estimated that 10 per cent of milk churns contained tubercle bacilli. Studies highlighted the fact that the 'public are exposed to the terrible risk of taking in the germs of consumption with their morning milk'.[35] By the Edwardian period, it had become customary to talk about milk and disease from a 'tuberculosis point of view'.[36]

the Danger to Man through the use as Food of the Meat and Milk of Tuberculosis Animals, part ii, minutes of evidence (London, 1898), p. 161.

[29] Harold Ernst, 'How Far May a Cow be Tuberculous Before her Milk Become Dangerous as an Article of Food?', *Practitioner* xlv (1890), pp. 146–60.

[30] 'Tuberculosis and Milk', *JCPT* xiii (1900), p. 242.

[31] 'The Tuberculosis Scare', *Dairyman's Record*, 20 Feb. 1900, p. 17.

[32] *Cowkeeper and Dairyman's Journal*, Mar. 1907, p. 193.

[33] John MacFadyean, 'Tubercle Bacilli in Cows' Milk', *JCPT* xiv (1901), p. 221.

[34] On the benefits of mixing, see Charles M. Aikman, *Milk: Its Nature and Composition: A Handbook on the Chemistry and Bacteriology of Milk, Butter and Cheese* (London, 1899).

[35] 'Consumption in Milk', *Veterinary Record*, 18 May 1901, p. 653.

[36] Arthur Littlejohn, 'Impure Milk', *Medical Officer*, 9 Mar. 1912, p. 102.

This and other evidence was used to support claims that tuberculous milk was responsible for the high incidence of bovine tuberculosis in children, which was felt to be damaging the health of the nation. Epidemiological and post-mortem studies point to the fact that whilst the danger to adults was minimal, the threat to children was considerable. These findings were confirmed by statistics from the Registrar General that highlighted how deaths from tuberculosis in adults, the 'meat-consuming population', had fallen whereas the death rate from the disease in infants, 'in the milk-consuming periods', showed no decrease.[37] Anxieties about degeneration and child health merged with the acceptance in medical and veterinary circles that milk was an important agent in the transmission of bovine tuberculosis. For many, the sale of diseased milk represented almost 'veiled infanticide'.[38] For the Association for the Prevention of Consumption the logic was obvious. It feared that because cows were susceptible to tuberculosis, 'sooner or later the bacilli find their way into the milk'. With infants and children under 5 consuming more milk than other age groups and offering fertile 'soil' for the bacilli, it concluded that their chance of infection was high.[39] The need for action as part of the crusade against consumption was clear. As A. Short, an Independent Labour Party councillor for Bootle explained, 'we are spending vast sums in building sanatoria for consumptives, we are holding great International Congresses to try and solve the problem of dealing with this fearful scourge effectually, and every town in the country is flooded with milk loaded with germs that spread the malady'.[40]

'All is muddle and confusion'

If contemporaries could agree that milk supplies were contaminated in some manner, and that controls were needed over production, problems emerged over the form these controls should take and who should be responsible. Farmers were repeatedly attacked for not producing clean or pure milk and, as Atkins has demonstrated, most of the radical suggestions for how to improve the milk supply came from outside the dairy industry.[41] Distinctions were made between pure milk from disease free herds and clean milk which was not adulterated or contaminated. A rhetoric of purity allowed a wide range of interest groups and campaigners to make pronouncements on the need for disease free milk. Numerous suggestions were put forward, including the

[37] John Penberthy, 'Veterinary Aspects of the Tuberculosis Problem', *JCPT* xx (1907), p. 296.
[38] 'Crusade against Consumption', *Sanitary Record*, 9 Mar. 1900, p. 212.
[39] 'National Movement for the Suppression of Tuberculosis', *ibid.*, 30 Dec. 1898, p. 693.
[40] A. Short, *A Municipal Milk Supply* (London, 1905), p. 7.
[41] Atkins, 'White Poison?', p. 210.

municipalisation of milk supplies to eliminate middlemen and to ensure hygienic production as part of moves to reduce infant mortality.[42] The heating of milk through sterilisation or pasteurisation to kill the bacilli was suggested and gathered momentum from late 1890s onwards in the wake of the second royal commission on tuberculosis. A number of MOsH distributed pamphlets in their towns, advising residents to boil their milk before use as part of local crusades against consumption.[43] Confidence was expressed in both methods to remove the threat of infection from bovine tuberculosis and hence avoid the need for expensive eradication measures, but as Peter Atkins has shown, sterilisation and pasteurisation remained controversial. For many both methods were seen as mere technological fixes that avoided the real problem of infection and did not strike at the 'root of the evil'. Their adoption therefore remained limited and problematic throughout the interwar period.[44] Equally controversial were recommendations for efficient herd management, regular inspection and the slaughter of diseased livestock. Farmers were advised regularly to test their milk, with some suggesting that they should furnish their customers with sanitary certificates.[45] Clean milk campaigns were launched by voluntary bodies including the National League for Physical Education and Improvement, and the National Health Society. They emphasised the importance of hygiene and the risk to children from diseased milk.[46]

Demand for pure or clean milk, despite the efforts of campaigners, remained low. For many, price and the amount of cream were the chief considerations, not purity. Before 1900, bottled or specially treated milk was a small and unprofitable part of the milk trade. Even by the 1940s, less than half the milk sold was pasteurised.[47] The need for public education as to the benefit of tubercle free milk was routinely called for, but public apathy was in part the result of ignorance and fear of tuberculosis combined with the higher cost of clean or pure milk. Clean milk was felt to be sold at a cost that 'renders it practically beyond the reach of the artisan class'.[48] Farmers and dairymen

[42] See Dwork, 'The Milk Option', pp. 51–69.

[43] Kensington Vestry, for example, informed the public as part of its crusade against consumption that all milk should be thoroughly boiled: Kensington Local Studies Library: Parish of St Mary Abbott's, Kensington Vestry minutes, 22 Mar. 1899.

[44] Atkins, 'Pasteurisation of England', pp. 37–51; TNA: PRO, Milk and Dairy Bill, 11 Feb. 1909, MH 80/3; J. Lloyd, 'Tuberculosis as Affecting Milk Supply', *Veterinary Record*, 9 Aug. 1902, p. 77.

[45] Charles Henry Leet, *Dairy Milk: Its Dangers and their Remedies* (Liverpool, 1896), pp. 9–10.

[46] National Health Society, *Milk Supply* (London, 1879); Joint Committee on Milk of the National Health Society and the National League for Physical Education and Improvement, *Milk Supply: Instructions for Ensuring the Supply of Clean Milk* (London, 1911).

[47] See E. H. Whetham, 'The London Milk Trade, 1860–1900', in *The Making of the Modern British Diet*, ed. Derek J. Oddy and Derek Miller (London, 1976), pp. 65–76.

[48] *Cowkeeper and Dairyman's Journal*, Jan. 1905, p. 531.

also resented the expense of producing clean or pure milk, and only the large dairies voluntarily improved methods of production. With demand low there was little incentive for small producers to make improvements. The publicity surrounding bovine tuberculosis in the wake of the second royal commission did encourage more people to start boiling milk, and saw greater efforts to improve production in the dairy industry. However, as James Beattie, the recently appointed professor of bacteriology at Liverpool University, told a conference on infant mortality in 1913, 'the energetic control of the milk supply was the only remedy'.[49]

How this 'energetic control' was to be achieved and by whom remained problematic. Whereas MOsH were keen to defend their role in inspecting meat, they were on less firm ground when it came to dairy cattle. They were therefore willing to cede some authority, but it was only following the second royal commission on tuberculosis that attempts were made to appoint qualified veterinary surgeons as dairy inspectors. By 1899, only thirty-eight local authorities had appointed a veterinary inspector.[50] Ad hoc arrangements were made that utilised the appointment of veterinarians under the Contagious Diseases (Animals) Acts and the extent of veterinary involvement in the inspection of cowsheds and dairies, and in the regulation of milk, was slowly extended.[51] The LGB increasingly came to favour the use of veterinarians in the inspection of all dairy cattle, reflecting a consensus among sanitary authorities and agriculturalists that veterinary examination was 'required'.[52] The need for veterinary inspection was confirmed by the 1909 Tuberculosis Order, which was introduced by the Board of Agriculture and Fisheries to limit the sale of milk from obviously tuberculous livestock. The Order conferred powers on veterinarians to inspect and test cattle suspected of having the disease. These powers were further confirmed by the 1914 Milk and Dairies Act which extended the system of veterinary inspection.[53]

However, the role given to veterinarians remained limited. Salaries were low: for example, in Epsom the veterinary inspector was paid £40 per annum, a salary on the same level as nuisance inspectors.[54] Many local authorities did not encourage regular inspection or call upon their veterinary officers to examine diseased livestock. Milk remained a medical question. 'Most veterinary officers', as Hardy notes, 'trod a fine line of professional demarcation in their responsibilities for dairy cattle'.[55] It was against these dilemmas over

[49] 'Conference on Infant Mortality', *Medical Officer*, 16 Aug. 1913, p. 81.

[50] TNA: PRO, Milk and Dairies Bill, MH 80/3.

[51] For example, Newport appointed a veterinary inspector in 1905 to examine milk entering the town: *BMJ* ii (1905), p. 1425.

[52] TNA: PRO, Power to Adrian, 1 June 1907, MH 80/3.

[53] *Lancet* ii (1909), p. 240.

[54] TNA: PRO, Milk and Dairies Bill, 1908, MH 80/3.

[55] Anne Hardy, 'Pioneers in the Victorian Provinces: Veterinarians, Public Health and the Urban Animal Economy', *Urban History* xxix (2002), p. 384.

responsibility and authority that attempts were made to regulate urban cowsheds and ensure a disease free milk supply.

The sanitary control of bovine tuberculosis

The primitive technological conditions of milk production in the mid nineteenth century required that for milk to be fresh on delivery it needed to be produced close to where it was to be sold. The result was an increasingly dense pattern of urban cowsheds. From the 1820s, these cowsheds attracted growing condemnation with the epizootic of rinderpest in the 1860s and endemic levels of bovine tuberculosis concentrating attention on the shocking conditions under which urban dairy cows were kept. Dairy cows were often kept 'huddled together in a space that does not allow them sufficient breathing room' in overcrowded and insanitary conditions.[56] It was argued that 'few greater nuisances exist' than urban 'cow-yards' which served 'as a *foci* of disease' and represented 'pest-houses of the most abominable nature'.[57] As Atkins has noted, from the 'human health point of view' these urban cowsheds represented an 'environmental hazard' that city administrators were keen to remove.[58]

Although new refrigeration and transport technology saw a shift in milk production from towns to rural areas, interest in urban dairy cows as 'hotbeds for the propagation of the tubercle bacillus' built on existing anxieties about urban cowsheds.[59] As the Board of Agriculture noted:

> since tuberculous animals excrete virulent material into the cowsheds mainly from the lungs and the bowels, and since they cannot be expected to make use of spittoons or other sanitary appliances of civilisation, the need for frequent cleaning and disinfection of cowsheds . . . is all the more pressing.[60]

That cowsheds and dairies represented 'hotbeds' of infection was easily explained by contemporaries who pointed to the warm, damp and overcrowded conditions under which cattle were frequently housed. In these conditions, 'It is no wonder then', wrote James King, chief veterinary inspector for Manchester, 'that many fine cows, newly purchased, and tied up in places with other cattle, which in many cases are suffering from tuberculosis, before

[56] James King, 'Tuberculosis and the Tuberculin Test', *JRAS* viii (1897), p. 320; *Annual Report of the Medical Officer of the Local Government Board* (London, 1878), p. 128; Henry Letheby, *Report to the Commissioners of Sewers on the Condition of the Cowhouses within the City: and on the quality of the milk supplied therefrom* (London, 1856).

[57] Hodson Rugg, *Observations*, p. 5.

[58] Atkins, 'White Poison?', p. 209.

[59] *JRAS* vi (1895), p. 657.

[60] *Journal of the Board of Agriculture* xx (1913–14), p. 478.

long become the subject of this disease'.[61] By targeting dairies and cowsheds through inclusive measures, it was hoped that a pure milk supply could be secured.

In attributing the spread of epizootic and zoonotic disease to urban cowsheds and dairies, a link was established between medical ideas about housing conditions, overcrowding and the spread of pulmonary tuberculosis, and debates on the conditions under which cattle were kept. Such an approach built on accepted notions that bad air and poor conditions weakened constitutions and initiated illness. Here a language was appropriated from the morally charged descriptions of the living and housing conditions of the urban poor. Overcrowding, the dangers of bad air, and fears about the increasing number of bodies in restricted urban spaces had become central to sanitary reform in the 1830s and 1840s. They formed a coherent language of description and concern that came to permeate the debates on public health and tuberculosis. Sanitary reformers, mainly ignorant about farming and dairy practices, were so used to talking about how the domestic and urban environment needed to be kept clean and pure that they adopted the same language when attacking cowsheds. Sanitary officials and veterinarians therefore readily claimed that overcrowded, poorly ventilated and badly constructed cowsheds and dairies were an ideal environment for the spread of bovine tuberculosis. In adopting this view of the cowshed and dairy as a polluted environment, bovine tuberculosis was placed within the same framework used to explain the spread of other contagious cattle diseases, notably rinderpest and pleuro-pneumonia. Better buildings and ventilation, linked to improved herd management, were therefore considered essential. This support for sanitary cowsheds built on a discourse that emphasised the natural germicide properties of light and air in the prevention of disease. By the 1890s, these criticisms of urban cowsheds had merged with the association preventive authorities were making between tuberculosis, dirt and squalor.[62]

Despite the encouragement given to improve and license urban cowsheds by the MAMOH and its successor, the SMOH, the regulation of dairies proved difficult to secure.[63] Local attempts were made to apply pressure on farmers and dairymen to improve conditions, but it was not until 1885 that the Dairies, Cowsheds and Milkshops Order granted local authorities greater regulatory powers.[64] The 1855 Nuisance Removal Act had proved ineffective in controlling milk production: sanitary authorities could only intervene when animals were kept under conditions 'to occasion nuisance or injury to public

[61] King, 'Tuberculosis', p. 320.

[62] Anne Hardy, *The Epidemic Streets: Infectious Disease and the Rise of Preventive Medicine, 1856–1900* (Oxford, 1993), p. 263.

[63] Wellcome: SMOH minutes of the general meeting, 19 Oct. 1859, SA/SMO/B.1.1.

[64] For example, under the 1862 Metropolis Local Management Amendment Act all cowsheds in London were made subject to annual licensing: *Lancet* ii (1862), p. 508.

health' and the difficulties of proving this had been considerable.[65] The 1885 Order offered greater scope for intervention. It was designed to supplement the Contagious Diseases (Animals) Act to ensure the registration of all those involved in the milk trade, to protect milk from contamination or infection, and to secure the inspection of dairy cattle. Under the Order, local authorities were expected to ensure that milch cows were kept under 'hygienic' conditions, particularly in relation to the construction of cowsheds and dairies to ensure proper ventilation, lighting, cleansing, and drainage. [66]

Campaigners sought to use this permissive legislation to clean up cowsheds in their attempts to prevent infection from milk-borne diseases. A desire to tackle the scourge of bovine tuberculosis increasingly came to underlie this action. MOsH regularly offered advice to farmers under the Order on the conditions in which cattle should be kept, drawing on the same language employed when discussing housing reform. Adequate ventilation was considered crucial. Detailed plans were drawn up for cowsheds that emphasised the need for hygiene, air and light, though much was left to the discretion of individual sanitary authorities.[67]

However, requirements varied between districts and regulation remained erratic. Attempts by urban councils to impose regulations, as in Leeds, frequently met with resistance from rural parishes who resented interference from urban sanitary officials. Districts also ignored the Order and in rural districts it was 'generally speaking, a dead letter'.[68] Even where the Order was applied, local authorities lacked sufficient powers to enforce the regulations. The size of rural districts and the sheer number of dairies and cowsheds created problems: premises could not always be inspected, especially if they were in isolated locations. When inspections did occur, they were often perfunctory.[69] Most MOsH and veterinary surgeons avoided demanding that structural alterations be made, aware that they were either unenforceable, impractical given local conditions, or would place a considerable financial burden on the occupier at a time when milk prices were falling. When inspectors were also veterinarians, the tendency was to make 'friendly suggestions' for fear of antagonising local farmers and prospective clients.[70] This was not a problem unique

[65] Edward Ballard, 'On Some Sanitary Aspects of Cowkeeping, and the Trade in Milk in London', *Milk Journal*, 2 Jan. 1871, p. 6.

[66] *Sixteenth Annual Report of the Local Government Board, 1886–7* (London, 1887), pp. 38–41.

[67] Edward F. Willoughby, *The Health Officer's Pocket Book: A Guide to Sanitary Practice and Law* (London, 1893), pp. 130–2; Gladstone Mayall, *Cows, Cow-Houses and Milk* (London, 1918), pp. 54–68.

[68] Dodd, *Milk Supply*, p. 49.

[69] *Annual Report of the Medical Officer of Health for Leicester County Council* (Leicester, 1901), p. 29.

[70] 'Sanitary Inspectors and the Supply of Milk', *Sanitary Inspectors' Journal* vi (1900–1), p. 5.

to the inspection of dairies: contemporaries readily assumed that all aspects of rural public health were at the mercy of landed and vested interests. Farmers were part of local social and political elites and hence were able to exert significant influence. Inspectors admitted that they frequently could do little; that their advice was regularly ignored, and in some cases 'the door of the outside cowshed [is] locked against us'.[71] Farmers complained that the regulations were too restrictive and that the Order was 'irksome and arduous'. Most objected to the cost of making improvements, refused to make changes, or moved to a neighbouring district where controls were less strict.[72] This created serious obstacles to reform.

By 1905 it was still believed that 'the great majority of cowsheds are much smaller than they should be, though this defect is being gradually remedied, still the conditions of very many leaves much to be desired'.[73] In Staffordshire, for example, despite the efforts of the MOH, most cowsheds were reported to be in a 'grossly filthy condition', overcrowded and with poor lighting and ventilation. Similar conditions were reported elsewhere.[74] Too much relied on the farmer or dairyman for the orders regulating cowsheds and dairies to be effective. Thomas Gibson, in a report for Wakefield city council, voiced a sentiment expressed by many MOsH and veterinary surgeons when he explained in 1914 that 'unless we get dairymen and farmers' to improve the sanitary conditions of their cowsheds 'we need never hope to eradicate tuberculosis from our cattle'.[75]

'The food that needs most careful protection'

The need to regulate cowsheds and dairies to prevent the conditions under which tuberculosis was propagated was only one part of the equation. From the 1880s, public health officials and veterinarians campaigned for more exclusive measures to target infected animals or their produce. Attention therefore increasingly came to focus on inspection to stop milk from tuberculous cows from entering the market. Although the absence of a 'strong "public interest" campaign' for clean milk, at least until the 1920s, allowed central and local government regulation of milk to remain 'extremely weak', according to

[71] 'Cowsheds in Rural Districts', *BMJ* ii (1907), p. 1009; *Annual Report made to the Urban Sanitary Authority of the Borough of Leeds* (Leeds, 1909), p. 112.
[72] Gilbert Murray, *Suggestions on the Management and Feeding of the Dairy Cow* (London, 1895), p. 6.
[73] Harold Sessions, *Cattle Tuberculosis: A Practical Guide to the Agriculturist and Inspector* (London, 1905), p. 49.
[74] *Report of the Medical Officer of Health for Staffordshire County Council* (Stafford, 1901), p. 97; J. Nash, 'A Clean Milk Supply', *Journal of Preventive Medicine* xiii (1905), p. 334.
[75] 'Unhealthy Cowsheds and Tuberculosis', *Medical Officer*, 28 Mar. 1914, p. 168.

French and Phillips the main blame rests with the Board of Agriculture and the power of the farming lobby.[76] The growing significance of milk to the agricultural economy, as more and more farmers entered the milk trade, and failing prices, encouraged dairy producers to develop a close relationship with the Board. They adopted a protectionist strategy in response to the agricultural depression and sought minimal state regulation. Their campaigns for the control of milk to be under the authority of the Board rather than the LGB, which was increasingly willing to secure national milk legislation to protect the public, ensured that the regulation of milk was left to local authorities and permissive legislation.[77] Here farmers were able to thwart action, especially in rural districts, and use their political influence to limit legislation.

Although piecemeal measures to regulate cowsheds were introduced in the 1860s, it was not until 1878 that regulations for licensing urban dairies, cowsheds and milkshops were introduced alongside the Contagious Diseases (Animals) Act of the same year. Given the emerging debate over the role of milk in the transmission of disease, and mounting anxiety about epizootics, these regulations were in part designed to prevent the sale of milk from diseased cows.[78] Powers were extended under the 1885 Dairies, Cowsheds and Milkshops Order, which included provisions for inspecting dairy cattle and outlined the precautions to be taken by milk sellers against infection or contamination. However, if the 1885 Order granted local authorities greater powers to prevent the sale of milk 'implicated' in the spread of contagious disease, no reference was made to tuberculosis as it was not included in the Contagious Diseases (Animals) Act. Nor was bovine tuberculosis mentioned in the 1890 Infectious Diseases (Prevention) Act, which allowed MOsH to stop the sale of milk from dairies where it was believed an infectious disease was present. At the same time, farmers and dairymen were under no obligation to notify the MOH of a diseased cow, leaving detection largely to chance. As with other areas of public health work, much depended on the character of the MOH and his subordinates.

Despite complaints from the Royal Agricultural Society that local sanitary authorities were being too stringent in the application of the Order, provision remained uneven.[79] Like much early sanitary legislation, the 1885 Order was permissive: local authorities were not compelled to frame regulations and 'extraordinary variations' existed between those authorities that issued milk regulations. With the milk trade dominated by small producers, and with milk being an adjunct to most general shopkeepers, sanitary authorities found it difficult to enforce milk regulations in a market in which 'if infected milk comes

[76] Michael French and Jim Phillips, *Cheated Not Poisoned? Food Regulation in the United Kingdom, 1875–1938* (Manchester, 2000), p. 159.
[77] See Phillips and French, 'State Regulation and the Hazards of Milk', p. 373.
[78] GCA: Memorandum on the milk supply of Glasgow, DTC 14/2/5/25.
[79] *JRAS* viii (1897), pp. 774–81.

from any one farm it may go to one shop to-day, to another to-morrow'.[80] Critics felt that the 'sanitary supervision' of milk was 'something of a farce'.[81]

The biggest problems existed in rural districts. Railways, refrigeration technology and rinderpest, which had decimated urban dairies in the 1860s, encouraged a shift in production from urban dairies to rural areas, so that by the end of the nineteenth century rural districts supplied the bulk of milk consumed in towns. This created major problems in combating bovine tuberculosis. Whereas urban dairies were comparatively well regulated, this was not the case in rural districts. By 1907, a quarter of rural district councils had still not made any regulations under the 1885 Order; in another quarter, adoption was 'merely perfunctory'.[82] Rural districts were routinely criticised for siding with dairy farmers, fearful that regulation would 'ruin the small farmer' and 'cripple a growing industry', and were attacked for not acting. Prejudices were confirmed by bacteriological tests on milk, which highlighted how a high proportion of milk from rural areas showed signs of the tubercle bacillus.[83] A rural/urban divide ensured that measures to control urban supplies had little impact on the amount of tuberculous milk reaching towns.

MOsH were all too aware that the inspection of dairies was at best rudimentary; that local authorities lacked powers to prevent the sale of milk drawn from diseased cows. The systematic inspection of 'dairies, and cows, so as to ensure good milk, free from the germs of tuberculosis' was seen as the solution.[84] Medical and veterinary witnesses to the second royal commission on tuberculosis thus called for greater powers of inspection, including the examination of dairy herds and milk for signs of tuberculosis. The commissioners were convinced that action was essential, but noted that it was futile to restrict the sale of tuberculous milk in cities when there were no safeguards against its introduction from outside. The commissioners consequently recommended the systematic inspection of all dairies and cowsheds; power to be granted to MOsH to suspend the supply of milk from any cow suspected of being diseased; and argued for fines for supplying infected milk. This was ideally to be linked to a policy of slaughter of diseased animals.[85]

[80] Peter J. Atkins, 'The Retail Milk Trade in London, *c.*1790–1914', *EcHR* xxxiii (1980), p. 531; GCA: 'Report on certain recent outbreaks of enteric fever', LP1/18.

[81] John Lindsay, 'Existing and Prospective Legislation *re* Milk Supply', *Journal of Meat and Milk Hygiene* i (1911), p. 673.

[82] J. Brittlebank, 'The Collection, Distribution and Contamination of Milk', *Journal of the Sanitary Institute* xxiv (1903), pp. 716–29; TNA: PRO, Milk and Dairies Bill, MH 80/1.

[83] *Ibid.*, MH 80/3; 'Milk Supplies and Tubercular Disease amongst Children', *Medical Officer*, 3 Sept. 1910, p. 138.

[84] Holborn District, *Report of the Medical Officer of Health for the half-year Ending December 1895* (London, 1896), p. 8.

[85] *Royal Commission . . . into . . . Controlling the Danger to Man through the use as Food of the Meat and Milk of Tuberculosis Animals*, minutes of evidence, pp. 13–15.

The revised 1899 Dairies, Cowsheds and Milkshops Order which resulted from the recommendations did permit local authorities to prohibit the sale of milk from cows certified by veterinary surgeons to be suffering from tuberculosis of the udder and allowed sanitary officials to inspect dairies when 'infectious disease attributed to milk' was reported. The Order also barred the sale of milk from diseased cows from being mixed with other milk. However, efforts by an overworked LGB to modify the 1885 Order did not go far enough.[86] The revised Order did not include clinical symptoms and, by limiting action to the obvious signs of tuberculosis of the udder, cows with other symptoms of the disease were excluded. Action was predicated on MOsH or veterinary surgeons providing clinical evidence of infection, which took time, ensuring that infected milk continued to be sold until this proof was obtained. Even when infected livestock were identified, the Order contained no provision for the removal of tuberculous cows. Nor did it require farmers to notify the local sanitary authority of diseased cattle or provide for regular inspection. It therefore remained possible for dairymen and farmers to avoid detection. For the *Journal of Comparative Pathology and Therapeutics*, the revised Order might therefore 'never have been issued'.[87]

Sanitary authorities did respond to the 1899 Order with further regulations to close loopholes that allowed milk to be sold legally even if it contained evidence of bovine or human tubercle bacilli. Manchester set the pattern. Impressed by the fact that tuberculous cows' milk was one of the most probable sources of infection in children, and with high levels of tuberculosis mortality in the city, James Niven, the radical MOH for Manchester, assisted by Sheridan Delépine, director of the public health laboratory at Victoria University, and by James King, the veterinary officer, made the 'elimination of tuberculosis from herds' their goal.[88] To achieve this they worked to reduce the amount of tuberculous milk supplied to the city. Delépine had carried out studies on tuberculosis, whilst Niven was interested in the problem of food and disease and in Oldham had been one of the first MOsH to enforce simple preventive measures against tuberculosis. They convinced the Council of the need to act on milk, aware that 'the present arrangements for protecting the public have almost entirely failed'.[89] Although the proposed powers were toned down following complaints from the local Chamber of Commerce, under the 1899 Manchester Milk Clauses all dairymen were required to notify Niven of cases of tuberculosis of the udder. Samples of milk from railway stations and farms

[86] *Twenty-ninth Annual Report of the Local Government Board* (London, 1900), pp. 9–10, 15–19.

[87] 'Tuberculosis and Milk', *JCPT* xiii (1900), p. 241.

[88] *Annual Report of the Medical Officer of Health of the Local Government Board*, appendix B (1908–9); James Niven, 'The Administration of the Manchester Milk Clauses', *BMJ* ii (1901), p. 314.

[89] James Niven, *Observations on the History of Public Health Effort in Manchester* (Manchester, 1923), pp. 126–7; *idem*, *On the Improvement of the Milk Supply*, p. 9.

were sent to Delépine to determine if the bacillus was present. Cowsheds in and around Manchester were also inspected by King. Farms from which samples of milk were found to contain evidence of the bacillus were inspected.[90] These measures were designed to improve detection, prevent the sale of tuberculous milk, and discourage farmers from supplying milk from infected cows. Delépine could demonstrate that the approach did reduce the supply of tuberculous milk in Manchester, with levels of the bacilli falling from 17.2 per cent in 1897/8 to 5.1 per cent in 1909. It was felt that the controls also resulted in a steady reduction in infant mortality.[91]

The policies adopted in Manchester were widely praised, and other areas moved to introduce similar measures. By 1910, sixty-seven boroughs and twenty-four urban districts had included clauses in local acts to regulate milk production.[92] To help police the milk trade, sanitary authorities also started to enter into arrangements with medical schools and public health laboratories to test their milk, reflecting the position bacteriological and chemical laboratories had started to secure in public health work. In Holborn, for example, milk was sent to the Jenner Institute for analysis; in Guildford samples were sent to St Bartholomew's Hospital.[93] For Sturdy and Cooter, these moves helped secure the position of the laboratory as part of a technology of surveillance.[94] Sanitary authorities did use bacteriology as a practical tool to determine whether milk contained harmful bacteria and as a way of identifying epidemic foci. Bacteriology was used to get round the problems of identifying diseased cows rather than as an 'administrative way of knowing', and created what Latour has suggested was a tool for more localised sites for sanitary intervention.[95] In Sheffield, for example, regular milk samples were used to direct inspection. All mixed milk brought into the city was tested for bovine tuberculosis; all milk from dairies and cowsheds in Sheffield with more than twenty cows had control samples taken. If tuberculosis was detected, the farm was visited and the cows examined by a veterinarian. Herds found to have bovine tuberculosis were prevented from supplying milk to Sheffield, though in practice this was not easy to enforce.[96] Similar regulations were introduced

[90] 'Meat and Milk in a "Great Town"', *Journal of Meat and Milk Hygiene* i (1911), p. 514.

[91] Sheridan Delépine, 'A Probable Effect of Control of Milk-Supply upon Infantile Mortality from Tuberculosis', *JCPT* xxv (1912), p. 129; National Clean Milk Society, *Campaign for Clean Milk* (London, 1916), p. 20.

[92] 'A Report on the Milk Supply of Large Towns', *BMJ* i (1903), pp. 1033–6.

[93] 'The Examination of Milk', *Dairyman's Record*, 20 Feb. 1900, p. 2.

[94] Steve Sturdy and Roger Cooter, 'Science, Scientific Management, and the Transformation of Medicine in Britain c. 1870–1950', *History of Science* xxxvi (1998), p. 437.

[95] Bruno Latour, *The Pasteurisation of France*, transl. Alan Sheridan and John Law (London, 1988).

[96] *Annual Report on the Health of the City of Sheffield* (Sheffield, 1910), pp. 48–50.

in Oldham, Birmingham and Cardiff.[97] By 1913, it was assumed that most sanitary authorities were testing their milk bacteriologically.[98]

However, even in those areas where milk regulations were adopted, obstacles continued to be encountered. MOsH considered provisions to inspect milk 'cumbersome', especially as magistrates were not always willing to prosecute.[99] The extent of bovine tuberculosis in dairy cattle made concerted action problematic, expensive and liable to opposition. The nature of the urban milk trade ensured the survival of a multitude of small retailers and single-shop enterprises that proved hard to inspect. The large number of itinerant salesmen compounded the problem. With milk an 'adjunct' to the main business of many grocery and general stores, owners were 'unwilling to expend any great effort in preserving its quality or purity'.[100] 'Formalities and legal technicalities' robbed measures of their utility and further frustrated MOsH.[101] There were not always the staff to carry out regular inspection. Nor was it frequently possible to find evidence of bovine tuberculosis on farms, especially as farmers would delay inspection so that they could remove obviously diseased livestock. It was felt that this forced MOsH to resort to 'wretched subterfuges'.[102]

Efforts were made to encourage co-operation. Luton rural council in a pamphlet, *Clean Milk Production*, told farmers that 'the Council's Sanitary Inspector is your friend and advisor' and encouraged local dairymen to contact him.[103] However, new regulations were often resisted by local dairy associations, and farmers and dairymen tried to circumvent inspection and control. Already under pressure from falling prices in a market in which competition was high and milk plentiful, and with the level of foreign imports creating concern, the restrictions imposed by the orders were seen to be damaging to trade. The added vigilance of local authorities and the disruption inspection caused encouraged those involved in the dairy industry to conceal evidence of any disease in their cows. Dairymen were repeatedly accused of being anxious 'only to dodge the inspector and put their milk on the market with as little trouble and as small expenses as possible'.[104] Dairy farmers would regularly mix milk to avoid detection and many involved in the milk trade blatantly ignored orders from public health officials. Farmers claimed they misunderstood the

[97] 'Tuberculous Milk', *Sanitary Record*, 18 Aug. 1904, p. 171.

[98] TNA: PRO, Milk and Dairies Act, 1914, MAF 52/5.

[99] *Annual Report of the Medical Officer of Health on the Vital and Sanitary Condition of the Borough of St Pancras* (London, 1905), p. 86.

[100] See Atkins, 'Retail Milk Trade', pp. 522–37.

[101] Dodd, *Milk Supply*, pp. 50–1.

[102] 'Attacking the Dairy Trade', *Cowkeeper and Dairyman's Journal*, Mar. 1904, p. 201.

[103] Bedfordshire and Luton Archives: Luton Rural District Council Illustrated Guide No. 1, 'Clean Milk Production', Z 630/5/1.

[104] Arthur Short, 'The Municipalisation of the Milk Supply', *Journal of Preventive Medicine* xiii (1905), p. 360.

Orders, whilst the language used in them made it possible for them to escape prosecution.[105]

The combination of these problems ensured that most of the powers to prevent the sale of tuberculous milk remained unenforced. Looking back on the measures introduced in Manchester, Niven explained that 'it was never expected that these clauses would do more than give a measure of control'.[106] Sanitary officials and veterinarians therefore continued to demand more effective milk legislation, whilst dairy farmers objected to 'ghastly' milk clauses and 'harassing restrictions'.[107] Neither side was satisfied with the existing situation. Although some sanitary authorities tried not to unnecessarily irritate dealers and dairymen, the vigour adopted by others generated opposition from the local dairy trade, who felt constantly under attack. Calls were made for uniform standards and national legislation, shifting the emphasis from fraud and adulteration to disease and health. At the same time, attacks were launched on the dairy industry and support for milk regulation was fuelled by the Pure Food and Health Society, the National League for Physical Education and Improvement, and the National Health Society. Sanitary officials pressed for the right to inspect any dairy whether it was in their district or not, and to be able to do so at any time, without notice and not just when an infectious disease was suspected, aware that it was futile to tackle bovine tuberculosis purely at an urban level when diseased milk from cows in the country was widely available.[108] Veterinarians argued that existing measures were 'incomplete' and asserted the need for them to play a more prominent role in inspection.[109] Local authorities sent resolutions to the LGB pressing for greater powers to regulate the 'serious menace to the public health' that milk from tuberculous cows was believed to represent.[110]

The LGB responded against a background of mounting criticism of the dairy industry and in the face of obstructionist policies from the agricultural lobby. It sought to push through legislation on public health grounds and secure the systematic inspection of all dairy cattle. Most urban boroughs were in favour of tighter controls, aware that existing measures 'did not go far enough', were 'cumbrous' and involved 'delay which allows a serious risk of spread of infectious disease'. However, efforts in 1909 to pass a Milk and Dairies Bill proved stillborn. Attempts by the Liberal government to push through the 1909 people's budget, which intended to finance social reform by raising death

[105] 'Milk from Diseased Cows', *Sanitary Record*, 8 Apr. 1909, p. 309.

[106] Niven, *Observations*, p. 131.

[107] *Cowkeeper and Dairyman's Journal*, Jan. 1905, p. 521; *ibid.*, June 1913, p. 401.

[108] 'The Milk and Dairies Bill', *Sanitary Record*, 22 July 1909, p. 84; *Annual Report of the Medical Officer of Health on the Vital and Sanitary Condition of the Borough of St Pancras* (London, 1905), p. 87.

[109] 'Legislation Regarding Tuberculous Cows', *Medical Officer*, 9 July 1910, p. 25.

[110] GRO: Cardiff Borough Council health and port sanitary committee, 16 May 1911, BC/C/6/43.

duties and other taxes, dominated parliament and the bill was quickly sacrificed in the face of concerted opposition from the dairy community and moves to water down the bill from the Board of Agriculture. Looking back, John Burns at the LGB felt that not so much had been lost from abandoning the bill, as it had sent a clear signal to the agricultural and dairy industry 'that they must put their house in order'.[111] Local authorities also reacted and stepped up their efforts to bring milk under closer supervision.

Although pressure continued to grow for legislation on milk quality, it was the publication of the final report of the royal commission on tuberculosis in 1911, and the provisions for tuberculosis treatment under the 1911 National Insurance Act, that drew attention to the need for action. The commissioners asserted that 'measures for securing the prevention of ingestion of living bovine tubercle bacilli with milk would greatly reduce the number of cases' of tuberculosis in children.[112] With local authorities starting to develop their tuberculosis services, and with growing demand to extend controls on food purity, the absence of uniform milk regulations was felt to be 'seriously imped[ing] the efforts of local authorities [in] the prevention of tuberculosis'.[113] Milk was an easy target for reformers as it was easily identified as a threat to health, especially child health. The need to regulate milk was seen as 'another step in the National campaign which is being waged against tuberculosis', but an attempt to balance competing health and agricultural claims left few happy. The new Milk and Dairies Bill announced in 1913 only grappled 'with the evil of tuberculosis' in its fully developed form, ensuring that the bill was not 'so stringent as either to cripple the milk trade or to cause a serious rise in the price of milk'.[114] Effective lobbying by agricultural interests ensured that the bill was limited in scope, forcing the LGB to moderate its proposals. The result was a compromise that left few satisfied. Consequently, urban MOsH felt that it did not go far enough, while their counterparts in rural districts saw the bill as a threat to their autonomy.[115] Veterinarians worried that it would undermine their work to secure greater responsibility for milk inspection and confirm their subordinate position.[116]

Yet, despite the criticisms levelled at the Milk and Dairies Bill, the general feeling was that it was a 'distinct step in the right direction'.[117] So insistent were the attacks on the milk industry that the public had become 'obsessed with the idea that milk production should be legislated for'.[118] As the popular press continued to draw attention to the dangers of milk, the LGB therefore

[111] TNA: PRO, 'Milk Supply', Nov. 1907, MH 80/3.
[112] *Royal Commission into the Relations of Human and Animal Tuberculosis*, PP (1911) xliii.
[113] GRO: Cardiff health and port sanitary committee, 15 Oct. 1912, BC/C/6/44.
[114] TNA: PRO, Milk and Dairies Bill, 28 Mar. 1913, MH 80/4.
[115] 'Milk and Dairies Bill', *Medical Officer*, 22 Mar. 1913, p. 131.
[116] *Veterinary Record*, 25 Jan. 1913, p. 449; *ibid.*, 29 Mar. 1913, pp. 618–19.
[117] 'Milk and Dairies Bill', *BMJ* i (1913), p. 735.
[118] 'The New Milk Bill', *Cowkeeper and Dairyman's Journal*, Apr. 1913, p. 301.

introduced a revised version. Herbert Samuel, president of the LGB, when announcing the new bill, claimed that it was

> absurd that when we are engaged in a great national campaign against tuberculosis, and we are spending millions to deal with pulmonary tuberculosis, we should almost ignore one-fourth of the whole number of cases – namely the cases of tuberculosis of a non-pulmonary character among children

derived from milk. It was a sentiment that echoed those expressed by many MOsH frustrated by their inability to tackle the problem of tuberculous milk. Although the new bill contained some of the same weaknesses of earlier legislation, it was more successful at balancing farming and public health interests. Consequently, it had a more favourable reception, especially following Burns's warning to the milk trade that 'worse for you will follow' if the bill was diluted.[119] By being linked to the Board of Agriculture's 1913 Tuberculosis Order (see chapter 10), it provided a mechanism for control and compensation. Existing regulations and local powers were replaced by uniform standards of inspection and a system of veterinary inspection was created which, it was hoped, would create the 'administrative machinery' for preventing the supply of tuberculous milk.[120] However, with the Act receiving the Royal Assent in August 1914, the LGB allowed the outbreak of war to delay implementation. It was finally put into effect in 1921.

Milk, health and the state

The abortive 1914 Act revealed the depth of concern about tuberculous milk. Tensions between rural and urban districts, and between the dairy industry and campaigners, had hampered attempts to secure a milk supply free from tuberculosis. Improvements in milk quality before 1914 were limited. Progress had been made, but 'merely touched upon the fringe' of the problem.[121] With measures to control the sale of milk difficult to implement, and with local initiatives effectively unable to ensure a pure or clean supply of milk from healthy cows, the *BMJ* argued in 1913 that the 'sterilisation of the milk is the only prophylactic measure we possess if we wish to safeguard our children against acquiring bovine tuberculosis'.[122] During and after the First World War, clean milk became central to consumer politics, replacing Edwardian concerns about cheap bread and free trade as popular demands were made for state

[119] TNA: PRO, Deputation, 29 Apr. 1913, MH 80/4.
[120] TNA: PRO, Milk proposals, 2 Mar. 1914, MH 80/5.
[121] GCA: Report of the Veterinary Surgeon to the Corporation of the City of Glasgow, 1913–14, C2/1/14.
[122] 'The Control of Our Milk Supply', *BMJ* ii (1913), p. 519.

controls and more interventionist food policies. Fears about adulteration persisted, but the desire for clean milk was driven by a blend of nutritional science, maternalist ideas, consumer politics and efforts to increase milk consumption, and by fears about bovine tuberculosis and the bacteria lurking in milk. Although efforts in the interwar period were directed at state-funded schemes to promote disease-free herds, the answer was increasingly seen to lie in a combination of pasteurisation and eradication.

10

Eradication?

Meat, milk and bovine tuberculosis

From the mid 1890s growing criticism of the effectiveness of meat and milk inspection to protect an unsuspecting public from bovine tuberculosis prompted attention to shift in favour of measures to stamp out the disease in cattle. As the second royal commission on tuberculosis explained, 'all precautions against the communication of tuberculous disease to human beings . . . must be regarded as temporary and uncertain palliatives so long as no attempt is made to reduce the disease among the animals themselves.'[1] It was considered absurd to try to protect human beings by inspecting meat and milk whilst doing nothing to limit the disease in the animals. Doctors and veterinarians united behind the need to stamp out bovine tuberculosis in cattle. They considered the elimination of bovine tuberculosis as intimately related to the crusade against consumption and a preventive strategy that followed a traditional prescription for public health through the removal of the source of infection. Between 1898 and the outbreak of war in 1914 efforts to protect the public and campaigns to eradicate the disease in cattle went hand in hand.

Eradicating bovine tuberculosis

The confusion surrounding the nature of bovine tuberculosis and reported levels of infection in cattle had discouraged any serious discussion of eradication in the mid nineteenth century. The disease was endemic not epizootic and until the 1870s its contagious properties were doubted. Instead, attention had focused on meat and milk inspection to check infection. However, as the limitations of inspection became clear, these approaches seemed insufficient to safeguard the public. Evidence that levels of bovine tuberculosis in humans and cattle were rising combined with fears about national degeneration to heighten concern. From the 1890s, an alternative strategy of prevention was therefore put forward based around the idea of stamping out the disease in

[1] *Royal Commission Appointed to Inquire into the Administrative Proceedings for Controlling the Danger to Man through the use as Food of the Meat and Milk of Tuberculosis Animals* (London, 1898).

cows. Consumers had little part in these discussions, which were played out between doctors, veterinarians, government officials, and the meat trade.

The possibility that contagious cattle diseases could be stamped out underpinned mid Victorian attempts to limit epizootics. Efforts to eradicate rinderpest, pleuro-pneumonia and foot-and-mouth asserted the primacy of slaughter in tackling epizootics and became a 'consistent feature of public animal health strategies, in Western Europe . . . for some three centuries'.[2] They demonstrated the need for action and provided a model for intervention that worked. In Britain, the basis for eradication schemes was established through the 1869 Contagious Diseases (Animals) Act passed in the wake of the cattle plague. The Act and successive legislation created a Veterinary Department (as part of the Privy Council) and an inspectorate, and established a system of notification, compulsory slaughter and compensation. The stamping out policies for rinderpest were adapted and modified for other contagious cattle diseases. Although the system was beset with problems – for example, notification was passive and reliant on farmers – the Veterinary Department made the need for 'the detection of disease' and 'the concentration of restrictive measures on those centres of disease' central to its efforts to prevent epizootics.[3] Measures to eliminate bovine tuberculosis built on these initiatives. They were further fuelled by the development of tuberculin in 1890 as a diagnostic agent, which overcame the problems of detecting the disease in cattle.[4]

Support for stamping out policies focused on the need to include bovine tuberculosis within the Contagious Diseases (Animals) Acts. Public health officials took up the issue, frustrated that they had limited powers to prevent meat and milk from tuberculous cattle from being sold and, given deficiencies with inspection, could not guarantee that tuberculosis was detected and infected meat and milk condemned. Many doctors therefore came to agree with Sheridan Delépine, director of the public health laboratory at the Victoria University, Manchester that the only way progress could be made was through 'the removal of the actual sources of infection'.[5] This was a return to the traditional public health policies that had been employed to control epidemics. Many elite veterinarians agreed. Common ground was found with doctors over the need to stamp out bovine tuberculosis in cattle whereas in other areas veterinarians and sanitary officials had found it hard to see eye-to-eye over

[2] John R. Fisher, 'To Kill or not to Kill: The Eradication of Contagious Bovine Pleuro-pneumonia in Western Europe', *Medical History* xlvii (2003), p. 314.

[3] *Annual Report of the Veterinary Department of the Privy Council Office* (London, 1879), p. 4; see 'Murrain of Beasts', *Milk Journal and Farm Gazette*, 1 Aug. 1872, p. 212.

[4] For the science of tuberculin see Keir Waddington, 'To stamp out "so terrible a malady": Bovine Tuberculosis and Tuberculin Testing in Britain, 1890–1939', *Medical History* xlviii (2004), pp. 29–48.

[5] 'Eradication of Bovine Tuberculosis', *Medical Officer*, 11 Jan. 1913, p. 13.

how to limit infection. The *Veterinary Record* feared that until action was taken 'we shall not be able to say that the public is secure against the invasion of contagious disease, due to infectious flesh'.[6] This was the opinion reached by the 1888 Departmental Committee on pleuro-pneumonia. It recommended that the Contagious Diseases (Animals) Acts be modified to included bovine tuberculosis. Although it rejected the need for compulsory slaughter and notification, worried about the consequences for the farming industry, the committee hoped that this would do much 'to minimise a disease so dangerous alike to animals and to mankind'.[7]

The support expressed by the Departmental Committee intensified debate. However, although a number of comprehensive schemes were outlined, few contemporaries were initially prepared to argue for the slaughter of all tuberculous cattle. How to award compensation, and at what level, were questions that were difficult to resolve. Most campaigners were conscious that, given estimates that a third of the cattle in Britain were suffering from the disease, any full-scale eradication programme would be financially damaging for both the farming industry and the state. It was also felt that consumers would suffer: predictions were made that eradication would result in a milk famine and drive up the price of meat. The Veterinary Department also vacillated. Keen to defend agricultural interests, which exercised a strong influence over the Department and its successors, it wanted to avoid upsetting farmers who, the Department felt, would not be prepared 'to suffer the enormous loss and inconvenience which . . . would entail upon them'.[8] As a result, attempts to include bovine tuberculosis under the Contagious Diseases (Animals) Acts were initially sidelined in favour of less controversial local voluntary schemes. Farmers were therefore encouraged to take steps to eliminate the disease in their herds, as 'it does not appear safe under present conditions to rely chiefly upon ordinary sanitary measures for the purpose of controlling bovine tuberculosis'.[9]

Fewer problems were encountered in Europe. Here eradication schemes were quickly established and provided inspiration for British campaigners. Denmark set the pattern. It had an innovative dairy industry and was one of the first countries to set up a state-sponsored eradication scheme based on the recommendations of Bernhard Bang, professor of medicine in Copenhagen and veterinary adviser to the Danish government. He outlined a voluntary programme in which cattle reacting to tuberculin were quarantined to shield

[6] 'Tuberculosis – Its Prevention', *Veterinary Record*, 10 Feb. 1894, p. 438; *ibid.*, 10 Sept. 1898, p. 141.

[7] *Departmental Committee appointed to inquire into Pleuro-pneumonia and Tuberculosis in the United Kingdom*, PP (1888) xxxii, pp. xxv, xxvi.

[8] *Royal Commission . . . into . . . Controlling the Danger to Man through the use as Food of the Meat and Milk of Tuberculosis Animals*, minutes of evidence, p. 10.

[9] 'The Distribution of Bovine Tuberculosis', *Medical Officer*, 11 Feb. 1911, p. 72.

the rest of the herd from infection. Those cattle considered too badly diseased were to be immediately slaughtered. Bang suggested that testing should be provided by the state in return for the quarantine of diseased livestock. What Bang proposed was a classic system of isolation to prevent the train of transmission reinforced by tuberculin as a diagnostic agent, a programme that reflected the faith European governments placed in the value of quarantine to prevent the spread of disease. There were clear advantages for the state: the scheme was relatively inexpensive, required minimum intervention on its part, and fitted with existing policies to control the spread of epizootics through isolation and slaughter. Bang's approach was adopted by the Danish government in 1893. Under the state-sponsored scheme, funding was given to farmers to test their cattle with tuberculin on the condition that they isolated all reacting cattle. By 1904, 404,651 animals had been tested and reported levels of bovine tuberculosis fell.[10]

Other European states followed Denmark's lead. Norway established a gratuitous testing service in 1895 and all animals found to be suffering from tuberculosis were either isolated or slaughtered depending of the severity of the disease. Sweden and Holland set up comparable schemes in 1897.[11] Voluntary testing, or at least clinical examination, was crucial, though each country had slightly different responses to the issue of compensation and slaughter. In France only those cattle that were 'obviously tuberculous' were killed and only those cows that had been in contact with these animals were tested.[12] Success was mixed: in Holland, farmers resisted testing, whilst the voluntary nature of many eradication schemes ensured that they were only partially successful.[13] Bang's method worked slowly: it had to be followed for a number of years, and any relaxation in vigilance could see tuberculosis reappear in a herd. Re-infection was disheartening, causing some to abandon the method.[14] There were also problems in securing proper isolation and the scheme demanded a 'clear conception of the nature of tuberculosis' which was not always present.[15] Notwithstanding these problems, eradication along the lines proposed by Bang remained the preferred solution for tackling bovine tuberculosis adopted by many European states.

Bang's scheme was well received in Britain. Interest spanned those who, from a public health perspective, wanted to prevent the spread of infection from bovine tuberculosis, to those who sought to protect farmers and dairymen as a means of reducing the financial burden on farmers already suffering the effects

[10] *Journal of the Board of Agriculture* xiv (1907–9), p. 436.

[11] *Ibid.*, pp. 496–501.

[12] *JRAS* vi (1895), p. 662.

[13] *Journal of the Board of Agriculture* xv (1908–9), pp. 496–501.

[14] William G. Savage, 'The Prevention of Human Tuberculosis of Bovine Origin', *Public Health* xxv (1911/12), p. 132.

[15] *Journal of the Board of Agriculture* xv (1908–9), p. 496.

of the agricultural depression. Eradication gained qualified approval from the second royal commission on tuberculosis and was quickly incorporated into local crusades against consumption as sanitary officials and veterinarians, conscious of the limitations of inspection, outlined measures to stamp out bovine tuberculosis in cattle.[16] Advice literature to farmers also began to suggest that those cows identified as tubercular should be isolated, then fattened, and sold for meat based on the assumption that cooking rendered meat safe. This, it was felt, would minimise the financial loss to the farming industry.[17] It was acknowledged that the process of creating a disease-free herd would be a slow one, but it was felt that by slaughtering 'the chief spreaders' of the disease, 'the first real, firm step forward to oust the bugbear will have been made'.[18]

Attempts were made to eliminate the disease from individual herds: some farmers did voluntarily test their cattle, selling those animals that reacted. For example, Thomas Carmichael, MP for Midlothian and a local landowner, started using tuberculin in 1896. By removing all reacting animals he felt that he had 'completely arrested' the spread of the disease.[19] However, efforts to produce milk from tuberculin-tested herds were isolated mainly because they had a limited financial success. One large farmer in Midlothian lamented that he had 'lost customers through it', and that his yields had fallen causing him to operate his dairy at a loss. There was little demand for milk from tested herds.[20] Under these conditions, voluntary testing, although well publicised, proved expensive and hence attracted only limited support.

Efforts by local authorities to establish eradication schemes were also patchy. The 1894 Contagious Diseases (Animals) Act had effectively devolved responsibility for the elimination of epizootics to local authorities and attempts to stamp out bovine tuberculosis were an extension of these responsibilities. In Birmingham, for example, an eradication scheme was started after a fact-

[16] *Royal Commission . . . into . . . Controlling the Danger to Man through the use as Food of the Meat and Milk of Tuberculosis Animals*, pp. 16–19; 'An Anti-Tuberculosis Programme', *Sanitary Record*, 25 Sept. 1896, p. 270; James Long, 'Consumption in Cattle Conveyable to Man', *Nineteenth Century* xlii (1897), p. 586; GCA: Report of the Veterinary Surgeon to the Corporation of the City of Glasgow, 1909–10, C2/1/10.

[17] See Edward Willoughby, *Milk: Its Production and Uses* (London, 1903), pp. 100–2; Thomas Brown, *The Complete Modern Farrier: A Compendium of Veterinary Science and Practice* (Edinburgh, 1900), pp. 529–30; T. Morrison Legge and Harold Sessions, *Cattle Tuberculosis: A Practical Guide to the Farmer, Butcher, and Meat Inspector* (London, 1898), p. ix.

[18] Gladstone Mayall, *Cows, Cow-Houses and Milk* (London, 1918), p. 106.

[19] *Veterinary Journal* xli (1895), p. 60; *ibid.*, xlii (1896), pp. 142–4. Similar practices were adopted by other farmers, see Wilfred Buckley, *Farm Records and the Production of Clean Milk at Moundsmere* (London, 1917), pp. 53–4.

[20] *Royal Commission . . . into . . . Controlling the Danger to Man through the use as Food of the Meat and Milk of Tuberculosis Animals*, minutes of evidence, pp. 294, 297.

finding visit to Denmark.[21] Other sanitary authorities were reluctant to establish full-scale eradication schemes, however. Whereas a number did encourage the use of tuberculin, most were 'unable to screw up the courage, or perhaps unwilling to unbutton their pockets' when it came to eradication.[22] Compensation remained an expensive and contentious issue. In comparison, meat and milk inspection were relatively inexpensive, whilst some local authorities believed that compensation discouraged farmers from eradicating the disease. It was felt that the only realistic solution to these problems was a state-funded eradication programme. As the Northumberland and Durham Dairy and Tenant Farmers' Association argued, it 'must be apparent to every right-thinking person that only in one way will it [bovine tuberculosis] ever be successfully combated, that is by Government taking it in hand, scheduling it, and paying fair compensation'.[23] Only through a national system of slaughter and compensation was it felt that tuberculosis could be 'got rid of' in this country.[24]

However, neither the Board of Agriculture nor the LGB were at first prepared to endorse a national eradication programme, preferring to support voluntary schemes. This was despite growing pressure from sanitary authorities and societies who wanted to protect consumers, from butchers' associations interested in compensation, and from veterinary organisations eager to extend their public health role. Although the LGB did come to favour national milk legislation and felt that tuberculosis could 'be stamped out in animals more quickly than it was being stamped out in human beings', it considered eradication problematic. Moves to stamp out bovine tuberculosis therefore received only tentative support from the LGB, which was already struggling under financial restrictions and overwork.[25]

Opposition at the Board of Agriculture ran deeper. Thomas Elliot, secretary to the Board, noted that because of all the conflicting opinions it was difficult to know what to do.[26] It was not just at a local level that confusion over the best solution could result in inactivity. Although the discovery of the diagnostic properties of tuberculin increased pressure on the Board to act, it remained unwilling to force through any measure, especially as it was

[21] Brennan de Vine, 'The Eradication of Tuberculosis from Dairy Herds Supplying Milk to Birmingham', *Journal of Meat and Milk Hygiene* i (1911), pp. 138–41.

[22] TNA: PRO, Milk and Dairies Bill, MH 80/3; 'Tuberculosis in London', *Veterinary Record*, 25 Feb. 1893, p. 470.

[23] *Journal of Meat and Milk Hygiene* i (1911), p. 97.

[24] 'Compensation for the Destruction of Tuberculous Cattle', *Sanitary Record*, 26 Oct. 1900, p. 364; *Select Committee on the Tuberculosis (Animals) Compensation Bill* (London, 1904), p. 11.

[25] 'Milk Legislation', *BMJ* i (1910), pp. 956–7.

[26] *Royal Commission . . . into . . . Controlling the Danger to Man through the use as Food of the Meat and Milk of Tuberculosis Animals*, minutes of evidence, p. 2.

unconvinced by the value of tuberculin.[27] Because the disease was endemic, it argued that the eradication of bovine tuberculosis was far more problematic than schemes used to stamp out epizootics like pleuro-pneumonia. It therefore dismissed the methods employed by Bang as difficult and liable to place a heavy burden on the farming industry, having realised that regular testing would find 'out too much' and open up a 'regular hornet's nest'.[28] Evidence of high levels of infection led the Board to balk at the cost implications of an eradication or mass testing programme at a time when it was already struggling with the financial burden of compensating stockowners for those epizootic cattle diseases included under the Contagious Diseases (Animals) Acts. It felt that the degree of danger had to be balanced against the cost of carrying out measures to reduce or remove the disease. For the Board the cost was too great.[29] It therefore retreated behind the claim that bovine tuberculosis was a trade risk and not the responsibility of the state.

Under pressure, and accused of deliberately ignoring trade delegations, the Board of Agriculture and LGB were manoeuvred into a position where they had to do something. There was a general demand for active measures for the eradication of bovine tuberculosis in cattle, whilst the publication of the interim reports of the third royal commission on tuberculosis stimulated debate on compensation just as the wider question of compensation for all diseased livestock was being discussed by the agricultural community.[30] Concerns were expressed that compensation would lead to 'speculative trade in low grade animals', 'incessant wrangling' over whether cattle were diseased, and force up the price of beef, but without compensation many felt that eradication would be a dead letter.[31] In response to the royal commission's interim reports, the Board of Agriculture issued the 1909 Tuberculosis Order to limit the sale of milk from obviously tuberculous livestock. Convinced that public opinion was favourable to measures to check the spread of bovine tuberculosis, and worried about the impact of the disease on the dairy industry, the Board had decided to act. The Order depended chiefly on a clinical diagnosis by a veterinary surgeon aided, where necessary, by microscopic and chemical analysis. Under the Order, all cattle suffering from tuberculosis of the udder and those emaciated from tuberculosis were to be slaughtered. Both represented the disease in its advanced state and for critics this only touched 'the fringe of the

[27] National Veterinary Association, *Sixteenth General Meeting* (London, 1898), pp. 13–29.

[28] TNA: PRO, Conference between representatives of the LGB and Board of Agriculture and Fisheries, 17 Mar. 1909, MH 80/3.

[29] 'Prevalence of Tuberculosis among Cattle in Britain', *JCPT* v (1892), p. 56; 'Compensation for the Destruction of Tuberculous Cattle', *Sanitary Record*, 26 Oct. 1900, p. 364.

[30] See University of Reading, Rural History Centre: National Farmers Union minutes, NFU AD1/1.

[31] 'The Tuberculosis Compensation Bill', *Sanitary Record*, 21 Apr. 1904, p. 343; *Public Health* xiv (1904/5), pp. 580–1; *Tuberculosis (Animals) Compensation Bill*, p. 57.

problem'.[32] To safeguard the farming industry, and to encourage farmers to report cases, compensation of up to £2 per animal was awarded; the Board was conscious that farmers 'would do nothing' without compensation.[33] It anticipated that the Order would check the spread of the disease, but advised sanitary officials and veterinarians to avoid 'heroic measures', which it believed would be detrimental. It recommended caution and 'due regard to the extent to which the disease is believed to exist amongst cows'. The Board was playing safe in an attempt to shield the dairy industry. No attempt was made to insist on periodic inspection, and the Order relied on farmers supplying information to the local sanitary authority.[34]

The 1909 Order was never fully implemented. The Board of Agriculture had always intended the Order to work in conjunction with the Milk and Dairies Bill (see chapter 9), but when the bill was withdrawn the Order was suspended partly in response to opposition from local authorities, which resented having to shoulder the burden of compensation. As Arthur Newsholme, MOH for Brighton recognised, the control of bovine tuberculosis was 'barred by expense'.[35] The veterinary and public health community was dismayed. Although the 1909 Order was criticised as a timid attempt to stamp out bovine tuberculosis, which was exactly what the Board of Agriculture had intended, it was heralded as the first stage 'in the inevitable campaign against tuberculosis'.[36] Suspending the Order was considered a retrograde step, as it was felt that even the limited measures contained in the Order 'shall have gone a long way in stopping up one of the main sources of infection'.[37]

The Board of Agriculture issued a new Tuberculosis Order in 1913. The final report of the 1912 Departmental Committee, established to consider national policy on tuberculosis following the 1911 National Insurance Act, had called for a co-ordinated effort to eliminate tuberculosis as the best way to 'obtain complete security'. A concerted move to prevent infection from bovine tuberculosis was judged a crucial component of these efforts.[38] Measures established by local authorities were deemed insufficient and the committee therefore recommended a national programme of eradication as part of a broader preventive strategy. The Board of Agriculture was hence under increasing pressure to act. It was also more willing to do so now that it had come under the control of the interventionist Walter Runciman. The 1913 Order extended supervision, allowed local authorities to inspect the farms from which diseased

[32] National Clean Milk Society, *Campaign for Clean Milk* (London, 1916), p. 21.
[33] TNA: PRO, Conference, MH 80/3.
[34] *Public Health* xiv (1904/5), pp. 580–1.
[35] TNA: PRO, Newsholme to Provis, 20 Mar. 1909, MH 80/3.
[36] 'The Present Position in Regard to the Sale of Milk', *Sanitary Record*, 28 Oct. 1909, p. 414; 'Legislation for Milk and Tuberculosis', *Veterinary Record*, 19 June 1909, p. 845.
[37] TNA: PRO, Burns to Chancellor, 29 July 1912, MH 80/4.
[38] *Departmental Committee on Tuberculosis*, PP (1912–13) xlviii.

animals came, and simplified the basis of compensation, with Treasury funds made available to cover half the compensation awarded by local authorities.[39] Runciman hoped that the Order would see bovine tuberculosis stamped out over a four- to five-year period.[40]

The Order was considered a step in the right direction, but within months criticism began to be voiced that it did not go far enough. For the *Medical Officer*, the Order quickly proved itself 'barren' and the *Morning Post* noted that the valuation system adopted was too elaborate to be effective. Local dairy associations complained that the amount of compensation awarded was inadequate and that the Order was expensive to administer.[41] The LGB was also conscious of the Order's limitations. Food safety had moved up the agenda at the LGB following the creation of the Food Section, and it had modified its stance on eradication. Increasingly troubled by bovine tuberculosis and milk, and under pressure from all sides, it wanted to push through legislation to secure the systematic inspection of all dairy cattle. It therefore pressed the Board of Agriculture to prepare a new order. The Board responded with a revised scheme that increased the level of compensation payable to stock-owners, aware that inadequate payments ensured poor compliance. However, the Order was suspended on the outbreak of war in 1914 and further efforts to eradicate bovine tuberculosis were postponed. As the Departmental Committee on Tuberculosis noted, 'the ultimate eradication of animal tuberculosis' is not impossible to achieve, 'but is likely to be a slow process'.[42]

The suspension of the 1913 Order and the outbreak of war ended two decades of debate on how best to eliminate bovine tuberculosis from cattle. Eradication as a preventive strategy had attracted growing support that first relied on voluntary provision and increasingly came to focus on the need for state intervention. However, the practical implementation of eradication was dogged with problems. Many of these were not unique to attempts to eliminate bovine tuberculosis but were present in efforts to stamp out epizootics.[43] Stockowners and farmers consistently demonstrated a 'want of appreciation of the advantages of the stamping out system as a means of getting rid of contagious diseases'.[44] Compliance proved a persistent problem for the Veterinary Department and its successors. In all cases, the 'pecuniary aspect

[39] TNA: PRO, Application to the Treasury, T1/16096; *Journal of the Board of Agriculture* xix (1912–13), pp. 152–3; 1043.

[40] 'The County Councils' Association and the Tuberculosis Order', *Medical Officer*, 12 Apr. 1913, p. 166.

[41] 'Compensation for Tuberculous Cows', *ibid.*, 21 Feb. 1914, p. 93; 'Tuberculosis Order', *The Times*, 25 Apr. 1913, p. 8.

[42] *Cowkeeper and Dairyman's Journal*, Apr. 1913, p. 305.

[43] See Fisher, 'To Kill or not to Kill', pp. 314–31.

[44] *Annual Report of the Agricultural Department, Privy Council, on the Contagious Diseases, Inspection and Transit of Animals* (London, 1888), pp. 5–6.

of the question' remained problematic both for the state and for the farming community.[45]

However, schemes to eradicate bovine tuberculosis encountered specific problems that discouraged the inclusion of the disease under the Contagious Diseases (Animals) Acts. Difficulties with identifying tuberculous livestock, even after the development of tuberculin, fuelled fears that a large percentage of the national herd would have to be slaughtered. It was recognised that this would result in ruin hence discouraging the adoption of a full-scale eradication policy. In addition, farmers were aware of the problems of securing disease-free herds and the sacrifices that were required. Many continued to see bovine tuberculosis as hereditary and, with the disease so widespread in cattle, they were fatalistic about the chance of stamping it out. 'Very few of them' therefore 'made the slightest effort to check the spread of the disease'.[46] Farmers also disliked interference from inspectors and had a low opinion of veterinary surgeons. Doubts were expressed about the ability of veterinary surgeons to undertake tuberculin testing, especially in the decade following the discovery of the test when it was felt that 'none of them has ever applied it or even seen it applied'.[47] Because of the high incidence of the disease, stockowners were unwilling to use tuberculin 'on account of the heavy loss to which the anticipated result might subject them', fearing that their herds would be decimated.[48] These concerns mirrored ongoing uncertainty of the effectiveness of tuberculin. Farmers and stockowners, who were naturally conservative, therefore resisted using tuberculin, creating a barrier to any effective eradication programme.

Eradication was also expensive for individual farmers, many of whom were moving into dairying to escape the effects of the agricultural depression. Estimates placed the annual cost of testing at between 2s and 3s per cow, which was beyond many smallholders.[49] In addition, suitable buildings and extra labour were required, all of which imposed a further financial burden on the farmer at a time when many were trying to cut down their overheads. Reacting cattle also had to be removed and replaced with healthy stock, which was expensive.[50] With compensation not forthcoming until 1909, farmers considered eradication schemes uneconomic and cumbersome. A strong agricultural lobby translated this opposition into effective pressure on the LGB and the Board of Agriculture, frustrating attempts to develop national eradication programmes. Farmers were blamed for inaction and were used as scapegoats

45 *Ibid.* (London, 1887), p. 5.

46 John MacFadyean, 'Tubercle Bacilli in Cows' Milk', *JCPT* xiv (1901), p. 221.

47 'Tuberculosis in Milch Cows', *The Times*, 12 Oct. 1898, p. 4.

48 'The Danger of Tuberculous Milk Supply', *Medical Officer*, 2 July 1910, p. 11; *Veterinary Record*, 10 Aug. 1901, p. 91.

49 *Royal Commission . . . into . . . Controlling the Danger to Man through the use as Food of the Meat and Milk of Tuberculosis Animals*, p. 20.

50 *Ibid.*, minutes of evidence, p. 270.

to explain why eradication measures either failed to work, or would not work.

However, eradication was not just frustrated by the agricultural lobby. Limited support was found at both the LGB and the Board of Agriculture. Both were anxious about the heavy financial cost of eradication.[51] Doubts were also expressed about the utility of tuberculin and the effectiveness of eradication, especially when both the LGB and the Board were fully aware that any scheme would encounter staunch resistance from the farming lobby unless backed up by compensation, which neither felt they were in a position to pay. They therefore initially sought to duck responsibility. It was only in the Edwardian period that the two boards adopted a more defined policy, but confusion over who was responsible and over competing spheres of influence made concerted action problematic. However, the cost implications of eradication continued to pose problems. It was not until the interwar period that political will began to shift in favour of funding a full-scale programme. As the World Health Organisation was to recognise in the 1960s, the challenge was to find 'economically feasible administrative measures' for eradicating the disease.[52]

Meat, milk and bovine tuberculosis

By the outbreak of war in 1914 the intense discussions on whether bovine tuberculosis was transmissible to humans or not that had characterised the mid nineteenth century had been largely settled. Disagreements between veterinarians and doctors remained, especially over whether the disease was localised in the organs and who was best suited to protect the public, but whereas veterinarians had first identified bovine tuberculosis as a danger to human health in the 1870s, by the 1890s most were convinced of the unity of the bovine and human forms of the disease and that food was an important source of infection. So entrenched were these ideas that although it was acknowledged by the 1910s that the threat from the disease to human health had probably been exaggerated, many MOsH readily agreed that tuberculosis was 'frequently introduced into the system by swallowing'.[53] Most food prosecutions by this date had come to involve the sale of tuberculous meat or milk. Local measures to limit infection through meat inspection, municipal abattoirs, and the regulation of the milk supply were well established and continued with little change until the outbreak of war in 1939, even if their practical implementation remained beset with problems.

[51] *Ibid.*, p. 11.

[52] Cited in Stephen Kunitz, 'Explanations and Ideologies of Mortality Patterns', *Population and Development Review* xiii (1987), p. 382.

[53] *Report on the Health of the City of Liverpool during 1908* (Liverpool, 1909), p. 63.

Although anxiety over bovine tuberculosis had begun to ebb from a high point in the 1890s as attention increasingly came to focus on the role of sanatoria and dispensaries in the treatment of consumption, by the Edwardian period no comprehensive scheme for tackling tuberculosis could ignore the bovine form of the disease. Sanatoriums did offer anti-tuberculosis campaigners the opportunity for intervention in the management of the disease, but bovine tuberculosis remained the one area in which the 'captain of all those men of death' could be effectively dealt with at a time when it was felt that preventive medicine had not yet produced 'any marked effect on the prevalence of tuberculosis'.[54] Fears of national degeneration and anxiety over infant and child welfare reinforced alarm about the sale of tuberculous meat and above all milk, contributing to a sense that any delay in limiting infection was tantamount to a 'continued massacre of innocents'.[55] The outbreak of war did see controls suspended and early attempts at eradication derailed, but the interwar period brought renewed efforts to combat bovine tuberculosis. Although eradication proved problematic and pasteurisation unpopular, debate was over how these were to be achieved with minimum cost not over the threat from bovine tuberculosis.

However, the alarm that came to surround bovine tuberculosis and the activity that was directed at preventing infection first from meat and then through milk was disproportionate to the number of cases of the disease in humans. What mattered more was the apparent danger of the disease to public health. Like rabies, the threat from bovine tuberculosis existed more at a rhetorical than an epidemiological level. The disease served a purpose for veterinarians and MOsH, providing the former with a convincing case for why they should play a role in public health and the latter with an opportunity to reform the meat and milk trades. They found in bovine tuberculosis a concrete threat around which they could articulate growing disquiet about meat and disease and about the relationship between the diseases of animals and humans that had first emerged in the 1850s as evidence accumulated of the evils of the meat trade and as the number of epizootics in cattle rose. Whereas the health consequences of eating meat from diseased livestock were contested throughout the 1850s and 1860s, experimental studies that increasingly asserted that bovine tuberculosis was contagious and could cross the species barrier not only challenged the dominant view of tuberculosis as hereditary but also legitimised alarm about diseased meat.

The redefinition of bovine tuberculosis as a zoonotic disease served to transform diseased meat from a loosely defined to a tangible threat, so that by the 1880s the problem of unwholesome meat and health had come to centre

[54] Edward Squire, 'Some Points in Relation to Tuberculosis and its Prevention', *PRSM* iv (1911), p. 28.

[55] John Penberthy, 'The Veterinary Aspects of the Tuberculosis Problem', *JCPT* xx (1907), p. 300.

on the 'Tuberculosis Meat Question'.[56] A decade later, bovine tuberculosis had become the model zoonosis, even if a plurality of ideas continued to exist on the nature of the disease. A similar process was seen in anxieties about milk, so that by the 1890s tuberculous milk and diseased milk had become synonymous. In breaking the conceptual barrier that differentiated the diseases of animals from those of humans, bovine tuberculosis shaped etiological research into the production of disease in humans through the agency of food. Alarm about bovine tuberculosis and human health also saw interest in food move beyond questions of purity to include the problem of food as an agent of infection. At the same time, work on the disease, first as part of studies into the pathology of contagion and then around the nature of the tubercle bacillus, helped define germ theories of disease. It encouraged refinements to the science and practice of bacteriology and reinforced the idea that the pathology of animal diseases could no longer be wholly separated from those of humans.

Once the contagious properties of bovine tuberculosis had been accepted, first by elite veterinarians and then by MOsH, and meat and milk identified as agents of infection, the disease became a public health issue. Once defined in this way, it became essential to protect the public from the dangers represented by 'coughing' cows. It is important to remember that in the mid nineteenth century prescriptions for public health were not just a matter of sanitation or epidemic disease control that subordinated concerns about the individual and poverty to the need to regulate the urban environment. They also embraced a range of other activities that included food regulation and the control of diseased meat. These areas were not amenable to a gospel of cleanliness or personal responsibility and required a preventive strategy that had little to do with sanitation.

If fears of bovine tuberculosis were played out in the national veterinary and medical press, it was the patchwork of local sanitary authorities that reacted to the perceived dangers of the disease, not the General Board of Health or its successors. The traditional institutions of local government were a crucial arena in shaping the public health movement and this was reflected in concerns about tuberculous meat and milk.[57] Local authorities and their officials translated concerns about bovine tuberculosis into action to safeguard 'innocent' consumers and minimise the health risks of eating meat and drinking milk from tubercular cows. Existing sanitarian measures designed to prevent nuisances were used and an approach adopted that concentrated on inspection and removal; an approach that belonged to an ideology of identification and surveillance that sought to break the train of transmission. Bacteriology brought a new tool for identification rather than a new way of conceiving strategies for regulating the meat and milk trade. Faith was placed in the value

[56] 'Tuberculous Meat', *BMJ* i (1889), p. 1359.

[57] Christopher Hamlin, 'Muddling in Bumbledom: On the Enormity of Large Sanitary Improvements in Four British Towns, 1855–85', *Victorian Studies* xxxii (1988/9), pp. 55–83.

of better inspection, public slaughterhouses, and the regulation of dairies and cowsheds to protect the public, even if considerable disagreement emerged over who had the expertise to make decisions about meat and milk. As a result, the 'grave responsibility' of local authorities 'to protect the consumer' from food that was deemed liable to cause disease was reinforced and a system of inspection was established that defined local attempts to regulate the meat and milk trade.[58] Such was the faith in the system established, that 'the Britisher . . . takes it for granted that the law of the country provides for the inspection of all butchers meat intended for his food, and consequently does not deem it necessary to make further investigation in the matter'.[59]

What part the public played in shaping these concerns remains uncertain, however. Evidence of direct public involvement in insisting on meat and milk inspection is limited. Unlike other contagious diseases, fears of bovine tuberculosis were essentially fashioned by elite veterinarians and doctors who defined the problem, drove debate and lamented that the public were not more interested in the threat they believed the disease represented. It was felt that public apathy created a barrier to effective measures to control the meat and milk trade by ensuring a ready market for cheap meat and milk from tuberculous livestock. This apathy was considered part of a general indifference to food quality. However, although apprehension about bovine tuberculosis and diseased meat was always greater among veterinarians and sanitary officials, the public, and especially the middle-class public, were increasingly worried about the disease. Public unease about food hygiene and questions surrounding pure food did grow in the period. Although greater anxiety was expressed about the threat from tuberculosis and milk than from tuberculous meat, growing public interest in food safety put pressure on local authorities to extend meat and milk inspection, whilst evidence from Stockport also demonstrates that the public did become more willing to take meat to inspectors if they had doubts about its quality.[60] This helped encourage the LGB and Board of Agriculture to pay more attention to the need to regulate the meat and milk trade and stamp out bovine tuberculosis in cattle. By the Edwardian period, public opinion was in favour of concerted measures to check the spread of bovine tuberculosis as an integral part of the crusade against consumption.

The alarm expressed by veterinarians and doctors, and the unease voiced by the public, were not matched by effective measures to regulate the meat and milk trade. Although some could point to a decline in cases of bovine tuberculosis in the human population, that the *Veterinary Journal* could still insist in 1911 that 'we must stop the sale of all tuberculous food' was indicative

[58] James Russell, *A Popular Exposition of the Modern Doctrine of Tuberculosis* (Glasgow, 1896), pp. 11–12.
[59] 'Meat Branding and Uniformity of Inspection', *Journal of Meat and Milk Hygiene* i (1911), p. 441.
[60] 'Selling Bad Meat in Stockport', *Sanitary Inspectors' Journal* ii (1895–6), pp. 316–17.

of how much remained to be done.[61] Historians have suggested that social intervention played an important role in mortality decline, but the history of meat and milk inspection indicates that not all areas of public health work progressed at the same rate, or were equally successful. Despite the improvements identified by public health historians, problems remained in translating concern and the science into practical action. Responses to bovine tuberculosis reveal the day-to-day problems encountered with inspection and the ongoing overt and covert opposition to local attempts to improve health. It also highlights how the regulation of food remained not only fraught with practical problems but also contested. The history of measures to prevent infection from bovine tuberculosis underlines the importance of trade interests in shaping food regulation and the role played by these groups in public health reform. Concerted rather than corporatist opposition from the meat trade and the agricultural lobby did limit attempts to improve inspection and stamp out bovine tuberculosis. At a local level, it was not unknown for farmers and butchers to make covert threats, implying that the lucrative cattle trade and dairy industry might be driven elsewhere if restrictive measures were adopted. The condemnation of diseased meat often relied on the support of local butchers, and the power of meat and dairy interests often determined the effectiveness of inspection.

However, problems with meat and milk inspection ran deeper than the extent of opposition, or the degree of support from local butchers or farmers. Public health was complicated by competing claims to expertise and disagreements over the nature of specific problems. Strong regional and urban differences prevailed, reflecting local physical, social, cultural or economic factors. This was clear in the inspection of meat and milk as emphasised by the 'dead meat drama' enacted in Glasgow in 1889. Fiscal and human resources were limited and could not address all the public health problems facing Victorian and Edwardian towns. The amount of work required was often beyond the resources available or the competence of the inspectors. Although improvements were made to the quality of inspection as standards and training were raised, gaps remained. It was often difficult to visit every site where nuisances were believed to occur. When it came to meat and milk inspection, the situation was further complicated by the sharp division that existed between urban and rural areas that was felt to reflect the backward nature of rural public health, an area that needs further investigation.

Responses to bovine tuberculosis not only emphasise local variations in public health and the complex questions that came to surround meat and milk, but they also confirm how it was not just the MOH who shaped local sanitary reform. MOsH did play a vital role in influencing the pace of public health reform but other sanitary officials, with their own professional interests, had

[61] 'Infantile and Bovine Tuberculosis', *Lancet* ii (1903), p. 788; *The Times*, 5 Aug. 1912, p. 4; *Veterinary Journal*, Apr. 1911, p. 196.

an important part in inspecting meat and milk, officials that have traditionally received little historical attention. At this level, medicalisation was slow to take effect, a phenomenon that raises questions regarding the impact of medical ideas on public health at a day-to-day level and the degree to which public health was professionalised. Competing claims for expertise were put forward and the authority of the MOH in determining whether meat was safe was frequently questioned, not only by other local sanitary officials, but also by lay committees and magistrates. Nor was it just doctors who defined public health problems. Bovine tuberculosis highlights how other professional groups, in this case veterinarians, were also important in advancing concerns about public health, and how doctors were not always first to identify risks to health. Although MOsH remained unwilling to cede responsibility for determining whether meat was diseased or not, their authority in doing so was increasingly questioned.

Despite the manifold problems in preventing infection from bovine tuberculosis through meat and milk, and the difficulties met with in eradicating the disease in cattle, by the interwar period the situation seemed more encouraging. Improvements to meat and milk inspection in the 1920s and 1930s, and further debate on eradication, extended the controls established in the late nineteenth century. Although the Second World War once more ensured that concerns about food were directed elsewhere, by the 1960s the prospect that bovine tuberculosis would no longer be a threat to public health had been largely realised. Levels of the disease in the national herd were in decline. The widespread adoption of pasteurisation was believed to have rendered milk safe and effective controls of slaughterhouses and a coherent system of meat inspection was felt to have done the same for meat. Bovine tuberculosis and other meat- and milk-borne diseases no longer appeared to be a threat to the consumer or to the agricultural industry. However, the emergence of new livestock diseases in the 1970s showed that the hazards of meat and milk had not gone away, a problem brought dramatically home by the furore that came to surround BSE.

The 1997 inquiry into BSE concluded that the effective control of animal diseases needed the type of surveillance regime that had been implemented by local authorities to deal with bovine tuberculosis.[62] Changes to the structure of public health controls and responsibilities had already been diluted before the BSE crisis. The implementation of the National Health Service in 1948 had undermined the public health profession. A bonfire of regulations in the 1980s, and Whitehall's belittlement of local government responsibility for overseeing slaughterhouses, allied to a series of funding cuts, made matters worse. The result was a curtailing of systems to prevent the spread of animal diseases that had been established in response to bovine tuberculosis in the late nineteenth century. The mechanisms put in place were far from perfect

[62] *The BSE Inquiry. Volume 1: Findings and Conclusions* (London, 2000).

and remained riddled with problems, but by the outbreak of war in 1914 bovine tuberculosis had redefined the problem of diseased meat, become the hidden enemy in the milk pail, and confirmed the risk from animal diseases to human health. Although the disease was not the Victorian and Edwardian equivalent of BSE, responses to it demonstrate that food safety and diseased meat are not new obsessions of the late twentieth century. An understanding of the disease reveals how public health concerns about meat have a longer history, and how fears about animal diseases have played an important role in shaping public health responses to the regulation of meat and milk. If the nature of food policy and concerns about food have shifted since the late nineteenth century, anxieties about meat and disease, animals and human health, have proved remarkably enduring.

Select bibliography

UNPUBLISHED PRIMARY SOURCES

Archives and Manuscripts, Wellcome Library for the Understanding of Medicine, 18 Euston Road, London

Society of Medical Officers of Health (SA/SMO)

Bedfordshire and Luton Archives and Record Service, County Hall, Cauldwell Street, Bedford

Luton Rural District Council Illustrated Guide No. 1, 'Clean Milk Production', Z/630/5/1

Corporation of London Record Office (CLRO), Guildhall, London

Abolition of slaughterhouses 1899–1927
Brief note on Meat Inspectors' office (Misc. Mss 366.14)
Cattle Markets Committee minutes
Cattle Plague Committee minutes (Misc. Mss 292.1)
Central Market Committee minutes
Common Council minutes
Destruction of Bad Meat notices (PD 149.10)
Food Shops, Stores and Places in the City (PD 149.15)
Port of London Medical Officer of Health reports
Port of London Sanitary Committee minutes
Prosecutions for Bad Meat (Misc. Mss 372.3)
Public Health Committee minutes
Sanitary Committee minutes
Sanitary Committee reports

Derbyshire Record Office, New Street, Matlock

Sessions role, Oct 1 Geo III, 1761 (D3551/4/72)

East Riding of Yorkshire Archives and Records Office, The Chapel, Lord Roberts Road, Beverley

Memorial under 'Act for the Registering and Securing Charitable Donations', 1813 (QDC/1/29)

Glamorgan Record Office, Glamorgan Building, King Edward VII Avenue, Cardiff

Cardiff Borough Council (BC)
Cardiff Local Board of Health minutes (BC L/B/1/1–4)

Glasgow City Archive, Mitchell Library, Glasgow

Complaints about the Cleanliness of Slaughterhouses (A2/1/3/12)
Corporation of Glasgow minutes, 1895–1914
Corporation of Glasgow reports (C2/1/5–14)
Diseases of Animals Acts (DTC 8/11)
Glasgow Committee of Health (E1/20)
Glasgow Markets Trust (MP26.464)
Glasgow Town Council minutes
Joint Subcommittee on Inspection of Dead Meat (MP29.168)
Markets and Slaughterhouses Department (DMK)
Medical Officer reports (LP1/18)
Medical Officer Statistical reports (DTC/14/2/37–41)
Memorandum as to the Inspection of Meat in Glasgow (MP20.601)
Memorandum on the Milk Supply of Glasgow (DTC 14/2/5/25)
Report to the Board of Supervision on Tuberculosis (Bovine) (MP18.58)
Report on the Subject of Tuberculosis in Cattle (MP17.314a)
Royal Commission on Market Rights and Tolls (MP20.593)
Special Subcommittee on Diseased Meat (MP9.24)
Tuberculosis in Cattle (MP17.301)

Hertfordshire Record Office, County Hall, Hertford

Manor and Borough of Kington, Court Rolls (D65/61)

London Metropolitan Archive (LMA), Northampton Road, Clerkenwell, London

Committee of the Society for Supplying the Poor with Meat Soup (P93/CTC1/55)

London County Council minutes
Shirley Murphy, 'Inspection of Milk and Meat', Nov. 1898 (LCC/MIN/10031)

The National Archive (TNA), Public Record Office (PRO), Kew

Medical Research Council (FD)
Ministry of Agriculture (MAF)
Ministry of Health (MH)
Treasury (T)

Oxford University, Department of Special Collections and Western Manuscripts, Bodleian Library

Papers of Christopher Addison

University of Reading, Rural History Centre

National Farmers Union (NFU)

OFFICIAL DOCUMENTS AND PUBLICATIONS (IN DATE ORDER)

Select Committee appointed to consider the Operation of the Acts for the Prevention of Infectious Disease in Cattle, PP (1850) xxiii
Report to the General Board of Health on a preliminary inquiry into the Sewerage, Drainage, and Supply of Water, and the sanitary condition of the inhabitants of Newton Heath (London, 1852)
Report from the Select Committee on the Adulteration of Food, Drink and Drugs, PP (1856) viii
Royal Commission on the Cattle Plague, PP (1866) xxii
Departmental Committee appointed to inquire into Pleuro-pneumonia and Tuberculosis in the United Kingdom, PP (1888) xxxii
Select Committee on Marking of Foreign Meat, PP (1893/4) xii
Royal Commission Appointed to Inquire into the Effect of Food Derived from Tuberculous Animals on Human Health (London, 1895)
Report of the Inland Sanitary Survey, 1893–95, submitted by the Medical Officer of the Local Government Board (London, 1896)
Royal Commission Appointed to Inquire into the Administrative Proceedings for Controlling the Danger to Man through the use as Food of the Meat and Milk of Tuberculosis Animals, PP (1898) xlix

Inter-Departmental Committee on Physical Deterioration, PP (1904) xxx
First Interim Report, Royal Commission into the Relations of Human and Animal Tuberculosis, PP (1904) xxxix
Select Committee on the Tuberculosis (Animals) Compensation Bill (London, 1904)
Report and Special Report from the Select Committee on Housing of the Working Classes Acts Amendment Bill, PP (1906) ix
Royal Commission into the Relations of Human and Animal Tuberculosis, PP (1907) xxxviii
Second Interim Report, Royal Commission into the Relations of Human and Animal Tuberculosis, PP (1907) lvii
Report of the Director of the Veterinary Department under the Diseases of Animals Acts, etc . . . for 1909, PP (1910) vii
Final Report, Royal Commission Appointed to Inquire into the Relations of Human and Animal Tuberculosis, PP (1911) xliii
Departmental Committee on Tuberculosis, PP (1912–13) xlviii
The BSE Inquiry. Volume 1: Findings and Conclusions (London, 2000)

ANNUAL REPORTS

Annual Report of the Agricultural Department, Privy Council, on the Contagious Diseases, Inspection and Transit of Animals
Annual Report of the Chief Veterinary Inspector for the City of Sheffield
Annual Report on the Health, Sanitary Condition, etc. of the Borough of Brighton
Annual Report of the Local Government Board
Annual Report of the Medical Officer of Health to the Privy Council
Annual Report of the Medical Officer of the Local Government Board
Annual Report made to the Urban Sanitary Authority of the Borough of Leeds
Annual Report of the Veterinary Department of the Privy Council Office
Birmingham, *Annual Report of the Medical Officer of Health*
Caernarvonshire County Council, *Annual Report of the Medical Officer of Health*
Cheltenham, *Annual Report of the Medical Officer of Health*
Edmonton, *Annual Report of the Medical Officer of Health*
Exeter, *Annual Report of the Medical Officer of Health*
Glasgow, *Annual Report of the Medical Officer of Health*
Holborn, *Annual Report of the Medical Officer of Health*
Kingston upon Thames, *Annual Report of the Medical Officer of Health*
Lambeth, *Annual Report of the Medical Officer of Health*
Manchester, *Annual Report of the Medical Officer of Health*
Portsmouth, *Annual Report of the Medical Officer of Health*
Preston, *Annual Report of the Medical Officer of Health*
Renfrew, *Annual Report of the Medical Officer of Health*
Report of the Health of the City of Liverpool
Report on the Health of Salford

Report on the Health and Sanitary Condition of Sunderland
Report on the Operations of the Sanitary Department of the City of Glasgow
Report on the Sanitary Condition of Leicester
Roxburgh, *Annual Report of the Medical Officer of Health*
St Giles Vestry, *Annual Report of the Medical Officer of Health*
St Pancras, *Annual Report of the Medical Officer of Health*
Sheffield, *Annual Report of the Medical Officer of Health*
Staffordshire, *Annual Report of the Medical Officer of Health*
Warrington, *Annual Report of the Medical Officer of Health*

NEWSPAPERS AND PERIODICALS

British and Foreign Medico-Chirurgical Review
British Journal of Tuberculosis
British Medical Journal
Cardiff and Merthyr Guardian
Contemporary Review
Cowkeeper and Dairyman's Journal
Dairyman's Record
Edinburgh Medical Journal
Edinburgh Veterinary Review
Food and Health
Food Journal
Glasgow Herald
Glasgow Medical Journal
Journal of the Board of Agriculture
Journal of Comparative Pathology and Therapeutics
Journal of Experimental Medicine
Journal of Meat and Milk Hygiene
Journal of Preventive Medicine
Journal of Public Health and Sanitary Review
Journal of the Royal Agricultural Society
Journal of the Sanitary Institute
Journal of State Medicine
The Lancet
Liverpool Courier
London Medical Gazette
Meat and Provisions Trades' Review
Meat Trades' Journal and Cattle Salesman's Gazette
Medical Chronicle
Medical Officer
Medical Press and Circular
Medical Times
Medical Times and Gazette

Medico-Chirurgical Review
Merthyr Express
Milk Journal
Milk Journal and Farm Gazette
National Association for the Prevention of Tuberculosis Bulletin
Nature
Newcastle Chronicle
Nineteenth Century
Practitioner
Proceedings of the Royal Society of Medicine
Public Health
Sanitary Inspectors' Journal
Sanitary Journal
Sanitary Record
Saturday Review
Sunday Times
The Times
Transactions of the Association of American Physicians
Transactions of the Epidemiological Society
Transactions of the Highland Agricultural Society of Scotland
Transactions of the Liverpool Medical Institution
Transactions of the Sanitary Institute
Tubercle
Veterinarian
Veterinary Journal
Veterinary Record
Western Mail

PRE 1948 PRINTED BOOKS AND ARTICLES

Archiv für pathologische Anatomie und Physiologie und für klinische Medizin xliv (1868)

Bibby's Book on Milk. Section IV. *Bovine Tuberculosis: Cause, Cure and Eradication* (Liverpool, [1911])

Congrés pour l'Étude de la Tuberculose chez l'Homme et chez les Animax (Paris, 1889)

Health Lectures for the People (London, 1885)

The Municipalisation of Milk, Fabian Tract No. 90 (London, 1899)

ORDERS *Conceived and Published by the Lord Mayor and Aldermen of the City of London, concerning the Infection of the Plague* (London, 1665)

The philosophie, commonlie called, the morals vvritten by the learned philosopher Plutarch of Chæronea. Translated out of Greeke into English, and conferred with the Latine translations and the French, by Philemon Holland of Coventrie, Doctor in Physicke (1663)

Public and Private Slaughterhouses: Report addressed to the Society for Providing Sanitary and Humane Methods for Killing Animals for Food (London, 1882)

Report to the General Board of Health on a preliminary inquiry into the Sewerage, Drainage, and Supply of Water, and the sanitary condition of the inhabitants of Newton Heath (London, 1852)

Report of the Second International Congress on Alimentary Hygiene 2 vols (1910)

Transactions of the British Congress on Tuberculosis 4 vols (London, 1901)

Transactions of the International Congress on Hygiene and Demography 2 vols (London, 1891)

Tuberculous Meat: Proceedings at Trial under Petitions at the Instance of the Glasgow Local Authority against Hugh Couper and Charles Moore (Glasgow, 1889)

Advisory Council of Science and Industry Queensland Committee, *Report of Committee on Tuberculosis, with special reference to cattle and pigs* (Brisbane, 1917)

Aikman, Charles M., *Milk: Its Nature and Composition: A Handbook on the Chemistry and Bacteriology of Milk, Butter and Cheese* (London, 1899)

Allan, Francis, *Aids to Sanitary Science, for the Use of Candidates for Public Health Qualifications* (London, 1903)

Andrews, O., *Handbook of Public Health Laboratory Work and Food Inspection* (London, 1901)

Ballard, Edward, *On a Localised Outbreak of Typhoid Fever in Islington ... traced to the use of impure milk, etc* (London, 1871)

Bantock, George, *Modern Dictionary of Bacteriology* (London, 1899)

Behrend, Henry, *Cattle Tuberculosis and Tuberculous Meat* (London, 1893)

Bell, Albert E., *The Pasteurisation and Sterilisation of Milk* (London, 1899)

Bonham, A. E., *A Practical Guide to the Inspection of Meat and Foods* (Exeter, 1915)

Bowley, A. L., and A. Burnett-Hurst, *Livelihood and Poverty* (London, 1915)

Brown, Thomas, *The Complete Modern Farrier: A Compendium of Veterinary Science and Practice* (Edinburgh, 1900)

Buchan, William, *Dr Buchan's Domestic Medicine; Or a Treatise on the Prevention and Cure of Disease, by Regimen and Simple Medicine* (Newcastle, 1812)

Buckley, Wilfred, *Farm Records and the Production of Clean Milk at Moundsmere* (London, 1917)

Chadwick, Edwin, *Report on the Sanitary Condition of the Labouring Population of Great Britain* (London, 1842)

Cheyne, George, *The English Malady: Or, A Treatise of Nervous Diseases of All Kinds, as Spleen, Vapours, Lowness of Spirits, Hypochondriacal, and Hysterical Distempers, etc* (London, 1733)

—— *An Essay on Regimen* (London, 1740)

Clifford Allbutt, Thomas, and Humphry Davy Rolleston, ed., *A System of Medicine* (London, 1906)

Cobbold, T. Spencer, *The Parasites of Meat and Prepared Flesh Food* (London, 1884)

Committee on Public Health, *Report of the International Commission on the Control of Bovine Tuberculosis, etc* (Ottawa, 1910)
Conn, H. W., *Practical Dairy Bacteriology* (London, 1918)
Cowderoy, John T., *Notes on Meat and Food Inspection for Sanitary Inspectors* (Kidderminster, 1912)
Creighton, Charles, *Bovine Tuberculosis in Man: An Account of the Pathology of Suspected Cases* (London, 1881)
—— *Contributions to the Physiological Theory of Tuberculosis* (London, 1908)
Crookshank, Edgar, *Bacteriology and Infectious Diseases* (London, 1896)
Cullimore, Daniel H., *Consumption as a Contagious Disease: With its Treatment According to the New Views, to which is Prefixed a Translation of Professor Cohnheim's Pamphlet, "Die Tuberkulose vom Standpunkte der Infections-Lehre"* (London, 1880)
Davies, Margaret Llewelyn, ed., *Life as we have known it* (London, 1990)
Dodd, Frederick, *The Problem of the Milk Supply* (London, 1904)
Dunhill, Thomas, *Health of Towns: A Selection of Papers on Sanitary Reform* (London, 1848)
Farr, William, *Vital Statistics* (Metuchen, NJ, 1975)
Fleming, George, *Animal Plagues: Their History, Nature and Prevention* (London, 1871)
—— *A Manual of Veterinary Sanitary Science and Police* 2 vols (London, 1875)
—— *The Contagious Diseases of Animals; their influence on the ... Nation, and how they are to be combated, etc* (London, 1876)
Fox, Wilson, *On the Artificial Production of Tubercle in the Lower Animals, etc* (London, 1868)
Gamgee, John, *Diseased Meat sold in Edinburgh, and Meat Inspection, in connection with the Public Health, and with the Interests of Agriculture. A Letter to the ... Lord Provost of Edinburgh* (Edinburgh, 1857)
—— *The Diseases of Animals in Relation to Public Health and Prosperity* (Edinburgh, 1863)
Gamgee, Joseph Sampson, *Cattle Plague and Diseased Meat in their Relations with the Public Health and with the interests of Agriculture. A letter to . . . Sir G Grey* (London, 1857)
—— *Cattle Plague and Diseased Meat in their Relations with the Public Health and with the interests of Agriculture. A second letter to Sir G Grey* (London, 1857)
Greenhow, Headlam, *Report on Murrain in Horned Cattle, the Public Sale of Diseased Animals, and the Effects of Consumption of their Flesh on Human Health* (London, 1857)
Hart, Ernest, 'The Influence of Milk in Spreading Zymotic Disease', *Transactions of the International Medical Congress* iv (London, 1881), 491–544
Houston, A. C., *Report on the Bacteriological Examination of Milk* (London, 1905)
Howarth, William, *Meat Inspection Problems, with Special Reference to the Development of Recent Years* (London, 1918)

Joint Committee on Milk of the National Health Society and the National League for Physical Education and Improvement, *Milk Supply: Instructions for Ensuring the Supply of Clean Milk* (London, 1911)
Kemp, William, *A brief treatise of the nature, causes, signes, preservation from, and cure of the pestilence collected by William Kemp* (London, 1665)
Leet, Charles Henry, *Dairy Milk: Its Dangers and their Remedies* (Liverpool, 1896)
Leffingwell, Alfred, *American Meat, and its Influence upon the Public Health* (London, 1910)
Legge, T. Morrison, and Harold Sessions, *Cattle Tuberculosis: A Practical Guide to the Farmer, Butcher, and Meat Inspector* (London, 1898)
Leighton, Gerald, and Loudon Douglas, *The Meat Industry and Meat Inspection* 5 vols (London, 1905)
Letheby, Henry, *Report to the Commissioners of Sewers on the condition of the cowhouses within the City: and on the quality of the milk supplied therefrom* (London, 1856)
Littlejohn, Arthur, *Meat and its Inspection: A Practical Guide for Meat Inspectors, Students, and Medical Officers of Health* (London, 1911)
Lydtin, A., George Fleming and M. van Hertsen, *The Influence of Heredity and Contagion on the Propagation of Tuberculosis and the Prevention of Injurious Effects from Consumption of the Flesh and Milk of Tuberculous Animals* (London, 1883)
MacFadyean, John, *Tuberculosis as Regards Heredity in Causation and Elimination from Infected Herds* (London, 1911)
May, Edward, *Tenure of Office of Sanitary Inspectors* (Manchester, 1883)
Mayall, Gladstone, *Cows, Cow-Houses and Milk* (London, 1918)
Moussu, G., and A. W. Dollar, *Diseases of Cattle, Sheep, Goats and Swine* (London, 1905)
Murray, Gilbert, *Suggestions on the Management and Feeding of the Dairy Cow* (London, 1895)
National Clean Milk Society, *Investigation of Milk* (London, 1918)
National Health Society, *Milk Supply* (London, 1879)
National Veterinary Association, *Sixteenth General Meeting* (London, 1898)
Newman, George, *Report on the Milk Supply of Finsbury* (London, 1903)
—— *Infant Mortality: A Social Problem* (London, 1906)
Newsholme, Arthur, *Hygiene: A Manual of Personal and Public Health* (London, 1906)
—— *The Prevention of Tuberculosis* (London, 1908)
Niven, James, *On the Improvement of the Milk Supply of Manchester* (London, 1896)
—— *Observations on the History of Public Health Effort in Manchester* (Manchester, 1923)
Ostertag, Robert, *Handbook of Meat Inspection* (London, 1904)
Parkes, Louis C., *Jubilee Retrospect of the Royal Sanitary Institute, 1876–1926* (London, 1926)

Pember Reeves, Maud, *Round About a Pound a Week* (London, 1913)
Pereira, Jonathan, *A Treatise on Food and Diet* (London, 1843)
Redford, A., *A History of Local Government in Manchester* (London, 1939)
Reid, George, *Practical Sanitation: A Handbook for Sanitary Inspectors and Others Interested in Sanitation* (London, 1895)
Robertson, William, *Meat and Food Inspection* (London, 1908)
Royal Institute of Public Health, *Transactions of the Congress held in Aberdeen* (London, 1901)
Royal Statistical Society, *The Production and Consumption of Meat* (London, 1903)
Rugg, Hodson H., *Observations on London Milk, shewing its unhealthy character* (London, [1850])
Russell, James, *A Popular Exposition of the Modern Doctrine of Tuberculosis* (Glasgow, 1896)
Savage, William, *Milk and the Public Health* (London, 1912)
Seaton, E., *Public Health Reports of John Simon* 2 vols (London, 1887)
Sessions, Harold, *Cattle Tuberculosis: A Practical Guide to the Agriculturist and Inspector* (London, 1905)
Sheldon, J. Prince, *Live Stock in Health and Disease* (London, 1902)
Sherer, John, *Rural Life* (London, 1868)
Short, A., *A Municipal Milk Supply* (London, 1905)
Smee, Alfred, *Milk, Typhoid Fever, and Sewerage* (London, 1873)
Smith, E., *Handbook for Inspectors of Nuisances* (London, 1873)
Smith, William, *A Sure Guide in Sickness and Health, in the Choice of Food, and Use of Medicine* (London, 1776)
Southey, Reginald, *The Nature and Affinities of Tubercle* (London, 1867)
Spencer Cobbold, T., *The Parasites of Meat and Prepared Flesh Food* (London, 1884)
Swithinbank, Harold, and George Newman, *Bacteriology of Milk* (London, 1903)
Taylor, Albert, *The Sanitary Inspector's Handbook* (London, 1906)
Vacher, Francis, *What Diseases are Communicable to Man from Diseased Animals Used as Food* (London, 1881)
—— *The Food Inspector's Handbook* (London, 1893)
Walker, Obadiah, *Of education, especially of young gentlemen in two parts, the second impression with additions* (1673)
Wallace, Robert, *Farm Live Stock of Great Britain* (Edinburgh, 1907)
Walley, Thomas, *The Four Bovine Scourges: Pleuro-pneumonia, foot-and-mouth-disease, cattle plague, tubercle (scrofula)* (Edinburgh, 1879)
—— *A Practical Guide to Meat Inspection* (Edinburgh, 1890)
Whitelegge, B. Arthur, *Hygiene and Public Health* (London, 1894)
Williams, William, *Principles and Practice of Veterinary Medicine* (Edinburgh, 1874)
Willoughby, Edward F., *The Health Officer's Pocket Book: A Guide to Sanitary Practice and Law* (London, 1893)

—— *Milk: Its Production and Uses* (London, 1903)
Wilson, George, *Handbook of Hygiene* (Edinburgh, 1873)
Woodward, John, *The State of Physic: And of Disease; With an Inquiry into the Causes of the Late Increase in Them* (London, 1718)
Wylde, W., *The Inspection of Meat: A Guide and Instruction Book to Officers Supervising Contract-meat and to all Sanitary Inspectors* (London, 1890)
Wynter Blyth, Alexander, *The Composition of Cows' Milk in Health and Disease* (London, 1879)
—— *A Manual of Public Health* (London, 1890)

POST 1948 PRINTED BOOKS AND ARTICLES

Alter, Peter, *The Reluctant Patron: Science and the State in Great Britain, 1850–1920* transl. Angela Davies (Oxford, 1987)
Antunes, J. *et al.*, 'Tuberculose e Leite', *História Ciências Saúde: Manguinhos* ix (2002), 609–23
Atkins, Peter J., 'The Retail Milk Trade in London, c.1790–1914', *EcHR* xxxiii (1980), 522–37
—— 'White Poison? The Social Consequences of Milk Consumption, 1850–1930', *SHM* v (1992), 207–27
—— 'The Pasteurisation of England: The Science, Culture and Health Implications of Milk Processing, 1900–50', in *Food, Science, Policy and Regulation in the Twentieth Century, International and Comparative Perspectives*, ed. David F. Smith and Jim Phillips (London, 2000), 37–51
—— 'The Glasgow Case: Meat, Disease and Regulation, 1889–1924', *AHR* lii (2004), 161–82
Barnes, David S., *The Making of a Social Disease: Tuberculosis in Nineteenth Century France* (Berkeley, 1995)
Blaisdell, John D., 'To the pillory for putrid poultry: Meat Hygiene and the Medieval London Butchers, Poulterers and Fishmongers' Companies', *Veterinary History* ix (1997), 114–24
Bonner, Thomas N., *American Doctors and German Universities: A Chapter in International Intellectual Relations, 1870–1914* (Lincoln, Nebraska, 1963)
Bourke, Joanna, 'Housewifery in Working-Class England, 1860–1914', *Past and Present* cxliii (1994), 167–97
Bowler, Catherine, and Peter Brimblecombe, 'Control of Air Pollution in Manchester prior to the Public Health Act, 1875', *Environment and History* vi (2000), 71–98
Brimblecombe, Peter, 'The Emergence of the Sanitary Inspector in Victorian Britain', *Journal of the Royal Society for the Promotion of Health* cxxiii (2003), 124–31
Brock, Thomas D., *Robert Koch, a Life in Medicine and Bacteriology* (Madison, WI, 1988)

Brown, Kenneth D., 'John Burns at the Local Government Board: A Reassessment', *Journal of Social Policy* vi (1977), 157–70

Brunton, Deborah, 'Policy, Powers and Practice: The Public Response to Public Health in the Scottish City', in *Medicine, Health and the Public Sphere in Britain, 1600–2000*, ed. Steve Sturdy (London, 2002), 171–88

Bryder, Linda, *Below the Magic Mountain: A Social History of Tuberculosis in Twentieth-century Britain* (Oxford, 1988)

—— '"Not Always One and the Same Thing": Registration of Tuberculosis Deaths in Britain, 1900–50', *SHM* ix (1996), 253–65

Burnett, John, *Plenty and Want: A Social History of Diet in England from 1815 to the Present Day* (London, 1966)

Bynum, W. F., '"C'est un malade": Animal Models and Concepts of Human Diseases', *Journal of the History of Medicine and Allied Sciences* xlv (1990), 397–413

Carlin, Martha, 'Fast Food and Urban Living Standards in Medieval England', in *Food and Eating in Medieval Europe*, ed. Martha Carlin and Joel Rosenthal (London, 1998), 27–51

Carlin, Martha, and Joel Rosenthal, ed., *Food and Eating in Medieval Europe* (London, 1998)

Chalmers, A. K., ed., *Public Health Administration in Glasgow: A Memorial Volume of the Writings of James Burn Russell* (Glasgow, 1905)

Cooter, Roger, and Steve Sturdy, 'Science, Scientific Management, and the Transformation of Medicine in Britain, c.1870–1950', *History of Science* xxxvi (1998), 421–66

Cranefield, Paul F., *Science and Empire:East Coast Fever in Rhodesia and the Transvaal* (Cambridge, 1991)

Cummins, S. Lyle, 'Jean-Antoine Villemin', in *Science, Medicine, and History: Essays on the Evolution of Scientific Thought and Medical Practice written in Honour of Charles Singer*, ed. Edgar Underwood, 2 vols (London, 1953), vol. ii, 331–40

Davin, Anna, 'Loaves and Fishes: Food in Poor Households in Late-Nineteenth Century London', *History Workshop Journal* xli (1996), 167–92

Davis, John, *Reforming London: The London Government Problem, 1855–1900* (Oxford, 1988)

Dwork, Deborah, *War is good for babies and other young children: A History of the Infant and Child Welfare Movement in England, 1898–1918* (London, 1987)

—— 'The Milk Option: An Aspect of the History of the Infant Welfare Movement in England, 1898–1908', *Medical History* xxxi (1987), 51–69

Dyer, Christopher, 'Did the Peasants really Starve in Medieval England', in *Food and Eating in Medieval Europe*, ed. Martha Carlin and Joel Rosenthal (London, 1998), 53–71

Eyler, John M., *Sir Arthur Newsholme and State Medicine, 1885–1935* (Cambridge, 1997)

Feinstein, Charles H., 'Pessimism Perpetuated: Real Wages and the Standard of Living in Britain during and after the Industrial Revolution', *JEcH* xcviii (1998), 625–58

Feldberg, Georgina D., *Disease and Class: Tuberculosis and the Shaping of Modern North American Society* (New Brunswick, NJ, 1995)

Finlay, Mark R., 'Quackery and Cookery: Jutus von Liebig's Extract of Meat and the Theory of Nutrition in the Victorian Age', *BHM* lxvi (1992), 404–18

Fisher, John R., 'Professor Gamgee and the Farmers', *Veterinary History* i (1979–81), 47–63

—— 'Not Quite a Profession: The Aspirations of Veterinary Surgeons in England in the Mid-Nineteenth Century', *Historical Research* lxvi (1993), 284–302

—— 'The European Enlightenment, Political Economy and the Origins of the Veterinary Profession in Britain', *Argos* xii (1995), 45–51

—— 'Cattle Plagues Past and Present: The Mystery of Mad Cow Disease', *Journal of Contemporary History* xxxiii (1998), 215–28

—— 'To Kill or not to Kill: The Eradication of Contagious Bovine Pleuro-Pneumonia in Western Europe', *Medical History* xlvii (2003), 314–31

French, Michael, and Jim Phillips, *Cheated Not Poisoned? Food Regulation in the United Kingdom, 1875–1938* (Manchester, 2000)

French, Richard D., *Antivivisection and Medical Science in Victorian Society* (Princeton, 1975)

Gradmann, Christoph, 'Robert Koch and the Pressures of Scientific Research: Tuberculosis and Tuberculin', *Medical History* xlv (2001), 1–32

Guerrini, Anita, *Obesity and Depression in the Enlightenment: The Life and Times of George Cheyne* (Norman, Oklahoma, 2000)

Haddad, George E., 'Medicine and the Culture of Commemoration: Representing Robert Koch's Discovery of the Tubercle Bacillus', *Orisis* xiv (1999), 118–37

Hall, Sherwin A, 'The Cattle Plague of 1865', *Medical History* vi (1962), 45–58

—— 'John Gamgee and the Edinburgh New Veterinary College', *Veterinary Record* lxxvii (1965), 1237–41

Hamlin, Christopher, 'Providence and Putrefaction: Victorian Sanitarians and the Natural Theology of Health and Disease', *Victorian Studies* xxviii (1985), 381–411

—— 'Politics and Germ Theories in Victorian Britain: The Metropolitan Water Commissions of 1867–9 and 1892–3', in *Government and Expertise: Specialists, Administrators and Professionals, 1860–1919*, ed. Roy MacLeod (Cambridge, 1988), 110–27

—— 'Muddling in Bumbledom: On the Enormity of Large Sanitary Improvements in Four British Towns, 1855–85', *Victorian Studies* xxxii (1988/9), 55–83

—— 'State Medicine in Great Britain', in *The History of Public Health and the Modern State*, ed. Dorothy Porter (Amsterdam, 1994), 132–64

—— 'Public Sphere to Public Health: The Transformation of "Nuisance"', in *Medicine, Health and the Public Sphere, 1600–2000*, ed. Steve Sturdy (London, 2002), 189–204

Hardy, Anne, 'Public Health and Experts: The London Medical Officers of Health', in *Government and Expertise: Specialists, Administrators and Professionals, 1860–1919*, ed. Roy MacLeod (Cambridge, 1988), 128–42

—— *The Epidemic Streets: Infectious Disease and the Rise of Preventive Medicine, 1856–1900* (Oxford, 1993)

—— 'On the Cusp: Epidemiology and Bacteriology at the Local Government Board, 1890–1905', *Medical History* xlii (1998), 328–46

—— 'Food, Hygiene and the Laboratory: A Brief History of Food Poisoning in Britain, c. 1850–1950', *SHM* xii (1999), 293–311

—— 'Pioneers in the Victorian Provinces: Veterinarians, Public Health and the Urban Animal Economy', *Urban History* xxix (2002), 372–87

—— 'Animals, Disease and Man: Making Connections', *Perspectives in Biology and Medicine* xlvi (2003), 200–15

—— 'Professional Advantage and Public Health: British Veterinarians and State Veterinary Services, 1865–1939', *Twentieth Century British History* xiv (2003), 1–23

—— 'Priorities in the History of Public Health and Preventive Medicine to 1945', in *Public Health and Preventive Medicine 1800–2000: Knowledge, Co-operation and Conflict*, ed. Astri Andresen, Kari Tove Elvbakken and William H. Hubbard (Bergen, 2004), 11–24

Hietala, Marjatta, 'Hygiene and the Control of Food in Finnish Towns at the Turn of the Century: A Case Study from Helsinki', in *The Origins and Development of Food Policies in Europe*, ed. John Burnett and Derek J. Oddy (Leicester, 1994), 113–29

Hunt, E. H., and S. J. Pam, 'Prices and Structural Response in English Agriculture, 1873–96', *EcHR* i (1997), 477–505

Johnson, A., 'Animal Disease and Human Health', in *The Advancement of Veterinary Science: The Bicentenary Symposium Series*, vol. 1, *Veterinary Medicine Beyond 2000*, ed. A. R. Michell (Oxford, 1993), 169–78

Kamminga, Harmke, and Andrew Cunningham, ed., *The Science of Nutrition, 1840–1940* (Amsterdam, 1995)

Knapp, Vincent, 'The Democratisation of Meat and Protein in Late-Eighteenth and Nineteenth Century Europe', *The Historian* lix (1997), 541–51

Koolmees, Peter, 'Veterinary Inspection and Food Hygiene in the Twentieth Century', in *Food, Science, Policy and Regulation in the Twentieth Century, International and Comparative Perspectives*, ed. David F. Smith and Jim Phillips (London, 2000), 53–68

Kunitz, Stephen, 'Explanations and Ideologies of Mortality Patterns', *Population and Development Review* xiii (1987), 379–408

Latour, Bruno, *The Pasteurisation of France*, transl. Alan Sheridan and John Law (London, 1988)

McCuaig, Katherine, *The Weariness, the Fever, and the Fret: The Campaign against Tuberculosis in Canada, 1900–50* (Montreal, 1999)

McNamee, Betty, 'Trends in Meat Consumption', in *Our Changing Fare: Two Hundred Years of British Food Habits*, ed. T. C. Barker, J. C. McKenzie, and JohnYudkin (London, 1966), 77–93

MAFF, *Animal Health, 1865–1965* (London, 1965)

Maulitz, Russell, *Morbid Appearances: The Anatomy of Pathology in the Early Nineteenth Century* (Cambridge, 1987)

Miller, D., and J. Reilly, 'Making an Issue of Food Safety', in *Eating Agendas: Food and Nutrition as a Social Problem*, ed. D. Mauer and J. Sobal (New York, 1995), 305–36

Nelson, Michael, 'Social-class Trends in British Diets, 1860–1980', in *Food, Diet and Economic Change Past and Present*, ed. Catherine Geissler and Derek J. Oddy (Leicester, 1993), 101–20

Oddy, Derek J., 'A Nutritional Analysis of Historical Evidence: The Working Class Diet, 1880–1914', in *Making of the British Diet*, ed. Derek J. Oddy and Derek Miller (London, 1976), 214–31

—— *From Plain Fare to Fusion Food: British Diet from the 1890s to the 1990s* (Woodbridge, 2003)

Ogawa, Mariko, 'Uneasy Bedfellows: Science and Politics in the Refutation of Koch's Bacterial Theory of Cholera', *BHM* lxxiv (2000), 671–707

O'Rourke, Kevin, 'The European Grain Invasion, 1870–1913', *JEcH* lvii (1997), 775–801

Pattison, Iain, *The British Veterinary Profession, 1791–1948* (London, 1984)

Pennington, Carolyn, 'Tuberculosis', in *Health Care as Social History: The Glasgow Case*, ed. Olive Checkland and Margaret Lamb (Aberdeen, 1982)

Perren, Richard, *The Meat Trade in Britain, 1840–1914* (London, 1978)

—— 'The Retail and Wholesale Meat Trade, 1880–1939', in *Diet and Health in Modern Britain*, ed. Derek J. Oddy and Derek Miller (Beckenham, Kent, 1985), 46–65

Phillips, Jim, and Michael French, 'Adulteration and Food Law, 1899–1939', *Twentieth Century British History* ix (1998), 350–69

Porter, Dorothy, 'Stratification and its Discontents: Professionalisation and Conflict in the British Public Health Service, 1848–1914', in *A History of Public Health: Health that Mocks the Doctors' Rules*, ed. Elizabeth Fee and R. Acheson (Oxford, 1991), 83–113

Porter, Roy, 'Man, Animals and Medicine at the Time of the Founding of the Royal Veterinary College', in *The Advancement of Veterinary Science: The Bicentenary Symposium Series*, vol. iii, *History of the Health Professions: Parallels between Veterinary and Medical History*, ed. A. R. Michell (Oxford, 1993), 19–30

Roberts, Elizabeth, *A Woman's Place: An Oral History of Working-Class Women, 1890–1940* (Oxford, 1984)

Roebuck, Janet, *Urban Development in Nineteenth Century London: Lambeth, Battersea and Wandsworth, 1838–88* (London, 1979)

Rosenkrantz, Barbara G., 'The Trouble with Bovine Tuberculosis', *BHM* lix (1985), 155–75

—— 'Koch's Bacillus: Was there a Technological Fix?', in *The Prism of Science*, ed. Edna Ullmann-Margalit (Boston, 1986), 147–60

Ross, Ellen, *Love and Toil: Motherhood in Outcast London, 1870–1918* (New York, 1993)

Scola, Roger, *Feeding the Victorian City: The Food Supply of Manchester, 1770–1870* (Manchester, 1992)

Shapin, Steven, 'Trusting George Cheyne: Scientific Expertise, Common Sense and Moral Authority in Early Eighteenth-century Dietetic Medicine', *BHM* lxxvii (2003), 263–97

Smith, David, and Malcolm Nicolson, 'The "Glasgow School" of Paton, Findlay and Cathcart: Conservative Thought in Chemical Physiology, Nutrition and Public Health', *Social Studies of Science* xix (1989), 195–238

—— and ——, 'Nutrition, Education, Ignorance and Income', in *The Science of Nutrition, 1840–1940*, ed. Harmke Kamminga and Andrew Cunningham (Amsterdam, 1995), 288–318

Smith, F. B., *The Retreat of Tuberculosis, 1850–1950* (London, 1988)

Super, John C., 'Food and History', *Journal of Social History* xxxvi (2002), 165–78

Swabe, Joanna, *Animals, Disease and Human Society* (London, 1999)

Szreter, Simon, 'The Importance of Social Intervention in Britain's Mortality Decline c.1850–1914: A Reinterpretation of the Role of Public Health', *SHM* i (1988), 1–37

—— 'Healthy Government? Britain, c.1850–1950', *Historical Journal* xxxiv (1991), 491–503

Taylor, David, 'The English Dairy Industry, 1860–1930: The Need for Reassessment', *AHR* xxii (1974), 153–9

Trentmann, Frank, 'Bread, Milk, and Democracy in Modern Britain: Consumption and Citizenship in Twentieth-Century Britain', in *The Politics of Consumption*, ed. Martin Daunton and Matthew Hilton (Oxford, 2001), 129–63

Vernon, Keith, 'Pus, Sewage, Beer and Milk: Microbiology in Britain, 1870–1940', *History of Science* xxviii (1990), 289–325

Waddington, Keir, 'The Science of Cows: Meat, Bovine Tuberculosis and the British State, 1880–1911', *History of Science*, xxxix (2001), 355–81

—— 'To stamp out "so terrible a malady": Bovine Tuberculosis and Tuberculin Testing in Britain, 1890–1939', *Medical History* xlviii (2004), 29–48

Whetham, Edith, 'Livestock Prices in Britain, 1851–93', *AHR* xi (1963), 27–35

—— 'The London Milk Trade, 1900–30', in *The Making of the Modern British Diet*, ed. Derek J. Oddy and Derek Miller (London, 1976), 65–76

White, Brenda, 'Medical Police: Politics and Police', *Medical History* xxvii (1983), 407–22

Wilkinson, Lise, 'Glanders: Medicine and Veterinary Medicine in Common Pursuit of a Contagious Disease', *Medical History* xxv (1981), 363–84

—— *Animals and Disease: An Introduction to the History of Comparative Medicine* (Cambridge, 1992)

—— 'Zoonoses and the Development of Concepts of Contagion and Infection', in *The Advancement of Veterinary Science: The Bicentenary Symposium Series*, vol. iii, *History of the Health Professions: Parallels between Veterinary and Medical History*, ed. A. R. Michell (Oxford, 1993), 73–90

Williams, E. F., 'The Development of the Meat Industry', in *The Making of the Modern British Diet*, ed. Derek J. Oddy and Derek Miller (London, 1976), 44–57

Wohl, Anthony S., *Endangered Lives: Public Health in Victorian Britain* (London, 1983)

Worboys, Michael, 'Germ Theories of Disease and British Veterinary Medicine, 1860–90', *Medical History* xxxv (1991), 308–27

—— '"Killing and curing": Veterinarians, Medicine and Germs in Britain, 1860–1900', *Veterinary History* vii (1992), 53–71

—— *Spreading Germs: Disease Theories and Medical Practice in Britain, 1865–1900* (Cambridge, 2000)

—— 'From Heredity to Infection: Tuberculosis, 1870–1890', in *Heredity and Infection: The History of Disease Transmission*, ed. Jean-Paul Gaudillière and Ilana Löwy (London, 2001), 81–100.

Yoder, Jon A., *Upton Sinclair* (New York, 1975)

Young, James Harvey, *Pure Food: Securing the Federal Food and Drugs Act of 1906* (Princeton, 1989)

UNPUBLISHED THESES, BOOKS AND PAPERS

Cronje, Gillian, 'Pulmonary tuberculosis in England and Wales' (unpubl. PhD diss., LSE, 1990)

Jones, Susan, 'Placing Disease: Transnational Perspectives on the Bovine Tuberculosis Debate, 1901–14', paper presented at Society for the History of Medicine spring conference, Sheffield Hallam University, Mar. 2002

Shell, Anna, 'TB or not TB eradicated?' (unpubl. PhD diss., Cardiff)

Watkins, Dorothy E., 'The English Revolution in Social Medicine, 1889–1911' (unpubl. PhD diss., London, 1984)

Index

abattoirs, *see* slaughterhouses
Aberdeen, 1, 122
Aberystwyth, 143
Académie de Médecine, 33, 34, 53
acts of parliament: Adulteration Acts (1860, 1862 and 1872), 72, 77; Bakehouse Regulation Act (1865), 77; City of London Sewers Act (1851), 25; Contagious Diseases (Animals) Acts, 40, 50, 90, 101–2, 106–8, 109, 138–40, 147, 161, 164, 166, 176, 177, 179, 181, 183; Cruelty to Animals Act (1872), 43; Dairies, Cowsheds and Milkshops Order (1885), 163–7, 170–1; Dairies, Cowsheds and Milkshops Order (1899), 168–9; Food Adulteration Act (1860), 77; Foreign Animals Order (1896), 146; Foreign Meat Regulations (1908), 147; Glasgow Markets and Slaughterhouses Act (1865), 94; Glasgow Police Act (1866), 94, 95; Infectious Diseases (Prevention) Act (1890), 166; Manchester Milk Clauses (1899), 168–9; Market and Fairs Clauses Act (1847), 72, 93; Metropolitan Local Management Act (1855), 73; Metropolitan Local Management Amendment Act (1862), 163 n64; Metropolitan Market Act (1851), 25, 73; Milk and Dairies Act (1914), 161, 171–3, 182; National Insurance Act (1911), 129, 172, 182; Nuisances Removal Acts (1846–8), 72, 77; Nuisances Removal Act (1855), 72, 163; Nuisances Removal Act (1863), 72; Public Health Act (1848), 72, 76, 77; Public Health Act (1872), 75, 146; Public Health Act (1875), 29, 43, 74, 77, 81, 95–6, 132, 150; Public Health (Regulation as to Food) Act (1907), 146–7; Town Improvements Clauses Consolidation Acts (1847), 72; Tuberculosis Order (1909), 161, 182–3; Tuberculosis Order (1913), 140, 173, 183–4; Veterinary Surgeons Act (1881), 40; Workshop Regulations Act (1867), 77
Addenbrooke's Hospital, 44
adulteration, 3, 23, 26, 72–3, 77, 79, 84; of milk, 153–4, 156, 159–60, 171, 174
Africa, 115
agriculture, 14, 15, 38, 68; and Board of Agriculture, 101–2, 172, 182, 185; depression in, 63, 166, 178–9, 184; importance of, 103, 144, 153–4, 165, 166, 189; milk and, 153–6, 158, 160–2, 164, 172. *See also* dairy industry, farmers
America, 4, 119, 129; bovine tuberculosis in, 3, 120 n38, 123–4, 127, 145–6; diseased meat in, 3, 18, 26, 145–6; and Koch, 123–4, 127. *See also* Bureau of Animal Industry, Chicago
anatomy, 44; comparative, 24
animal disease: and contagion, 1–4, 6, 12–14, 20–49, 53–4, 58–60, 96–106, 109–10, 112–30, 154–9, 186–7; and Contagious Diseases (Animals) Acts, 40, 50, 102, 106–9, 138–40, 147, 161, 164, 166, 176, 181, 183; eradication of, 2, 4, 14, 38, 40, 49–50, 90, 92, 101–2, 106, 123, 160, 175–85, 190; fear of, 1–2, 4, 12, 14, 27; and germ theory, 7, 28, 30–1, 35, 39, 42, 46–51, 56, 93, 99, 119, 187; and LGB, 42–3, 48, 109, 119, 125, 141, 146, 168, 183, 188; and localised v. generalised infection, 58–62, 64, 68, 81, 84–5, 89–92,

97–101, 105–6, 117–18, 141–2, 158; and medical profession, 6–9, 19–29, 33–6, 41–5, 48–62, 64–9, 83–4, 139–40, 147, 186–7; and MOsH, 20–9, 43, 54, 56–8, 60, 73–4, 83–4, 139, 147, 155–6, 158–61, 186–90; parasitic, 38, 42, 55; pathologists and, 6, 30, 33, 36, 45, 47–8, 53–4, 157; and Privy Council, 35, 49–50, 64, 92, 104–5, 176–7, 183; relation with human health, 1–4, 6–7, 12–14, 20–69, 97–106, 108–10, 112–30, 139, 154–9, 160–1, 175, 185–91; research into, 4, 6–8, 21, 27–9, 30, 32–5, 39, 45, 48, 53–4, 103–6, 109–10, 112, 114–30; veterinarians and, 1, 7–8, 24–5, 32–3, 36–41, 43–4, 48, 53, 55, 59–60, 71, 105, 131, 137–9, 147, 161–2, 168, 185–90; virulence, 114–15, 121, 123, 126. *See also individual diseases*
anthrax, 17, 23, 25, 26, 32, 39, 46, 47, 52, 53, 60, 71
Arloing, Saturnin, 123
Armstrong, Henry, 86
Association of Veterinary Officers of Health, 137, 139
Australia, 123

bacteriologists, 49, 101, 114, 120, 123, 136
bacteriology: and bovine tuberculosis, 3, 7, 13, 22, 31, 46–51, 56, 62, 103–5, 109, 114–15, 119–26, 128, 135–6, 154, 157, 169–70, 187–8; and Koch, 7, 13, 31, 46–51, 56, 103, 114, 124–6; laboratories for, 48, 101, 103, 119–22, 128–9, 135–6, 169; and meat inspection, 101, 135–6, 139, 187; and medical profession, 46–9, 56, 62, 103–4, 129, 135–6, 139; methods, 46, 48, 103–4, 114, 119–22, 135, 187; and milk, 154, 157, 167, 168–70, 187; and MOsH, 48, 50, 56, 62, 74–5, 135–6, 139, 169–70, 187; opposition to, 62; and public health, 7, 47–51, 56, 62, 74–5, 103, 119, 128–9, 135–6, 139, 167, 169–70, 176, 187; and royal commissions on tuberculosis, 8, 103–5, 109, 119–26, 128; and state, 102–5, 109, 119–22, 128–9; and tubercle bacillus, 7, 31, 46–51, 56, 62, 101, 103, 113–14, 120–2, 126, 128, 135–6, 187; and veterinarians, 47–9, 103, 139. *See also* Robert Koch
bakers, 67
Ballard, Edward, 30
Bang, Bernhard, 177–9, 181
Barnes, W., 152
Battersea, 134
Beattie, James, 161
beef, *see* BSE, meat
Behrend, Henry, 62
Berlin, 31, 34, 45, 46, 53, 114, 124
Berlin Physiological Society, 31, 45, 46, 124
Berry, Justice, 97, 98–9, 100
Berry, William, 140
Billing, G., 132
Birkenhead, 19, 44, 88, 104
Birmingham, 139, 142, 170, 179–80
Bishop Auckland, 84–5
blood, 126; and diseased meat, 53, 57, 61–2, 96, 98, 123, 140–1
Blythwood farm, 119
Board of Agriculture: and bovine tuberculosis, 101–2, 104, 106, 110, 161, 166, 171–3, 180–3, 185, 188; and compensation, 101–2, 106, 110, 173, 180–3, 185; and Contagious Diseases (Animals) Acts, 102, 181; and dairy industry, 161–2, 166, 172–3, 181–2, 188; and diseased meat, 101, 188; and epizootics, 101–2, 176, 180–1; and farmers, 101–2, 166, 172–3, 181, 184–5; and food safety, 101–2, 161, 162, 166, 172–3, 180–3, 185, 188; and Glasgow trial, 101–2; and LGB, 161–2, 166, 172–3, 180–1, 183, 185, 188; and milk, 161–2, 166, 171–2, 181–3, 188; and tuberculin, 181, 185; and Tuberculosis Order, 161, 173, 181–3
Boer War, 156
Bond, William, 132
Bootle, 159
Bott, Gibson, 19, 20
Bouley, Henri, 47
bovine tuberculosis, *see* tuberculosis, bovine
Bradford, 134
Brighton, 57, 182; butchers in, 87 n72, 88, 89
Bristol, 73, 86
British Cholera Commission, 118–19
British congress on tuberculosis (1901), 8, 13, 99, 112–19, 123, 126, 141. *See also* John McFadyean, Robert Koch, royal commission on tuberculosis
British Medical Association (BMA), 44, 45

Brown Animal Sanatory Institution, 35, 41, 43, 104
BSE (bovine spongiform encephalopathy): and beef, 1–2; and CJD, 1; in cows, 1–4, 190–1; and public, 1–2; and state, 1–2, 190–1; and veterinarians, 1. *See also* CJD, food scares, meat
Buchanan, George, 101
Buchanan, George Seaton, 146
Bulmer, Thomas, 43
Bureau of Animal Industry, 123
Burns, John, 172, 173
Bury, 151
butchers: associations, 64, 71, 90, 94, 100, 106–8, 110, 139, 142, 151, 180; and bovine tuberculosis, 19, 50, 54, 56, 63–4, 82–3, 87–90, 94–7, 99–101, 105–8, 110, 117, 125, 131, 141–4, 148–52, 189; in Brighton, 87 n72, 88, 89; and compensation, 90, 100, 106–10, 125, 142, 180; and cooking, 65–8; criticism of, 64, 89–90; and diseased meat, 19–20, 54, 56, 63–4, 71, 82–4, 87–90, 94–7, 99–101, 105–8, 110, 117, 134–5, 139, 141–4, 148–52, 180, 188–9; dishonesty of, 19, 82–3, 88–90, 105, 142, 143–4, 148–9; expertise of, 75, 79–80, 84, 97, 99–100, 107–8, 134–5, 189; in Glasgow, 89, 94–7; Honourable Company of, 21, 71; influence of, 50, 84, 87–6, 94, 150, 189; juries, 84; and localised v. generalised infection, 58–62, 89–90, 98, 100, 141; in Manchester, 57, 151; and meat inspection, 64, 71, 79–80, 82–4, 87–90, 94, 100–1, 106–9, 117, 134–5, 139, 141–4, 148–52, 188–9; prosecution of, 26, 27, 29, 43, 56, 75, 87, 90, 92–3, 96–7, 99–101, 107–8, 125, 141–2, 144–5, 185; and royal commissions on tuberculosis, 105–8, 110; and seizure of meat, 53, 56, 64, 71, 82–3, 84, 87–90, 100, 106–7, 109, 125, 132–3, 142–5, 149, 189; in Sheffield, 64; and slaughterhouses, 82, 148–52; in Swansea, 139; voluntary surrender of meat, 87–8, 96, 143–4. *See also* National Federation of Butchers and Meat Traders' Associations

Calmette, Albert, 123
Cambridge, 44, 45
Cambridge University, 44
Canada, 123
Cardiff, 77 n26, 131, 136, 150, 170
Carmichael, Thomas, 179
cattle: and bovine tuberculosis, 2–3, 12–14, 17, 19, 30–1, 33, 36–40, 52, 54, 56, 59–62, 82, 90, 96, 99, 102, 104–6, 110, 114–15, 125, 128, 145–7, 157–9, 162, 168, 170, 175, 177–8, 186–7; BSE in, 1–4, 190–1; compensation and, 90, 100, 106–7, 108–9, 125, 173, 176, 180–5; and contagion, 1–3, 6, 7, 21, 23, 27–9, 30, 32, 28–9, 38–9, 53, 96–106, 109–10, 112–30, 154–9; dairy, 13, 154–9, 168; diseased, 1–4, 10, 12–14, 17, 19–20, 21, 23, 27–8, 32, 35, 37–40, 43, 52, 54, 56, 59–62, 90, 96, 99, 102, 104–6, 110, 114–15, 125, 145–7, 154–9, 162, 167–8, 175–8, 186–7, 190; epizootics and, 3, 17, 19, 21, 23, 25, 27–8, 30, 32, 35–6, 37, 38–9, 43, 52, 90, 92, 101–2, 106, 115, 138, 146, 162, 176, 178, 186; imported, 27, 28, 145–7; inspection of, 12, 108, 145–7, 161–74, 176–85; polio, 10; slaughter of, 17, 21, 23–4, 104, 108, 110, 146, 160, 167, 176, 178, 180, 181. *See also* cattle plague, pleuro-pneumonia, rinderpest, tuberculosis
cattle plague: 30, 37, 39, 42; commission on, 28–9, 35; impact of, 27–9, 38, 146, 162, 176. *See also* rinderpest
Chadwick, Edwin, 77
Chauveau, Jean-Baptiste, 53, 92
chemistry, 22, 181
Chicago, 3, 145–6, 147
children: and bovine tuberculosis, 3, 12, 14, 66, 124, 126, 153–4, 156–7, 159–60, 168, 172; and degeneration, 3, 9, 12, 14, 153, 156, 186; health of, 3, 5, 12, 14, 48, 153–4, 156–7, 159–60, 186; and milk, 5, 9, 12, 153–4, 156–7, 159–60, 169, 172
cholera, 118–19
Christison, Robert, 22
CJD (Creutzfeldt-Jakob Disease): and BSE, 1–2; and food scares, 1; and state, 1–2. *See also* BSE, food scares, meat
Clark, Andrew, 33
coffee, 154
Cohnheim, Julius, 34, 47
Coleman, Edward, 40
compensation: and Board of Agriculture, 101–2, 106, 110, 173, 180–3, 185; and bovine tuberculosis, 90, 100, 106–10,

125, 132, 142, 173, 180–5; and butchers, 90, 100, 106–10, 125, 132, 142, 176, 180; and Contagious Diseases (Animals) Acts, 90, 107–9, 176; for epizootics, 90, 101–2, 106, 176, 180–1; and farmers, 90, 101–2, 106–9, 173, 176, 180–5; and LGB, 109, 110, 125, 173, 180, 184–5; and meat inspection, 90, 106–9, 112, 125, 132, 142; and National Federation of Butchers and Meat Traders' Associations, 90, 106–8, 110, 142; opposition to, 90, 180–1; and royal commission on tuberculosis, 106–10, 117, 125, 181; and Tuberculosis Order, 173, 181–3; and veterinarians, 90
congresses: hygienic (1891), 30; international veterinary (1985), 57, 59; Paris (1867), 33; Paris (1888), 50, 57, 92, 97; on tuberculosis (1901), 8, 13, 99, 112–19, 123, 126, 141
consumption: food, 14–20, 23, 52, 154, 160–1, 176; of meat, 6, 15–20, 23, 52, 54; of milk, 154, 160–1, 174, 176. *See also* standard of living
consumption, pulmonary, *see* tuberculosis
contagion: and animal disease, 1, 2, 4, 6, 12–14, 20–49, 53–4, 58–60, 96–106, 109–10, 112–30, 154–9, 186–7; and bacteriology, 3, 31, 46–9, 62, 187; and blood, 53, 57, 61–2, 96, 98, 123, 140–1; and bovine tuberculosis, 3, 6–7, 13–14, 29–49, 53–8, 60–2, 96–106, 109–10, 112–30, 154–9, 186–7; and localised v. generalised disease, 58–62, 68, 81, 84–5, 141–2, 158; and meat, 1–3, 6–7, 14, 25, 28–9, 35–8, 43, 52–69, 96–106, 109–10, 131, 141–2, 157, 186–7; medical profession and, 27–9, 33–6, 41–9, 54, 60–2, 187; and milk, 5, 9, 11–12, 92, 106–8, 111–14, 116–17, 124–5, 128, 154–9; MOsH and, 26–7, 43, 48–50, 62; pathology of, 27, 29–36, 42–3, 53, 114, 187; and tuberculosis, 3, 21–2, 31–4, 36, 41–51, 59–62, 96–106, 109–10, 112–30, 186–7; veterinarians and, 24–5, 30, 32–3, 36–9, 43–4, 59–60, 138, 186
Contagious Diseases (Animals) Acts, 102, 106, 147; and Board of Agriculture, 102, 181; and bovine tuberculosis, 50, 90, 107–9, 138, 176; and butchers, 107–8; and compensation, 90, 107–9, 176; and Departmental Committee on Pleuro-pneumonia and Tuberculosis, 50, 177; and epizootics, 40, 101–2, 106, 138, 176–7, 179, 181; and meat inspection, 106–8, 138–9, 140; and milk inspection, 161, 164, 166; and veterinarians, 40, 138–40, 161
cooking: benefits of, 64–8, 154, 179; and housing, 66–7; of meat, 64–8, 154, 179
Copenhagen, 177
Coulton, William, 84–5
Couper, Hugh, 96–7
cowpox, 21, 39
cows, *see* cattle
cowsheds: and bovine tuberculosis, 3, 13, 55, 162–73, 188; conditions in, 3, 13, 55, 162–5; disease in, 3, 12, 55, 162–3, 188; hygiene, 154–5; inspection of, 162–73, 188; licensing of, 162–5; in Manchester, 168–9; and milk inspection, 162–73, 188; and MOsH, 162–5; and nuisances, 162–4; rural, 164, 167; urban, 3, 13, 162–5; and veterinarians, 162–3, 164–5
Creighton, Charles, 44–5
Crewe, 87
customs officials, 147

dairies: and bovine tuberculosis, 13, 157–9, 162–73, 188; conditions in, 13, 154–5, 162–3, 165; disease in, 13, 154–9, 162–3, 165–73; inspection of, 111, 161, 163–73, 185, 188; in Manchester, 168–9, 171; and MOsH, 163–8, 170–1, 173, 189; regulation of, 113, 161, 163–73, 188; rural, 162, 164–5, 167; urban, 13, 162, 167; and veterinarians, 160–1, 168, 172. *See also* milk
dairy industry: and Board of Agriculture, 161–2, 166, 172–3, 181–2, 188; and bovine tuberculosis, 113, 153–4, 158, 161–74, 179–80, 183, 189; and Contagious Diseases (Animals) Acts, 161, 164, 166; and disease, 13, 154–61, 165, 189; and eradication of bovine tuberculosis, 160, 179–81, 183; influence of, 160–1, 165–6, 181, 183, 189; and LGB, 161, 166, 168, 171–3; and milk inspection, 113, 161, 166, 168, 170, 172; regulation of, 113, 161–74, 189; and Tuberculosis Order, 161, 173, 183. *See also* agriculture, dairies, farmers, milk

degeneration: and Boer War, 156; and child health, 3, 9, 12, 14, 153, 156, 186; fears of, 5, 9, 12, 14, 108, 153, 156, 159, 175, 186; and milk, 9, 108, 153, 156, 159, 175, 186
Delépine, Sheridan, 168–9, 176
Denmark, 180; and bovine tuberculosis, 115, 177–8; and tuberculin, 178
Department of the Environment, Food and Rural Affairs, 2
Departmental Committee on Pleuro-pneumonia and Tuberculosis: appointment of, 49–50; and bovine tuberculosis, 50, 56, 61, 65, 92, 96–7, 102, 177, 183; and compensation, 177; and Contagious Diseases (Animals) Acts, 50, 177; and cooking, 65; and localised infection, 61
Departmental Committee on Tuberculosis, 129, 182, 183
Dick Veterinary College, Edinburgh, 36, 37, 66
diets, 15–23, 52, 54, 154; meat in, 6, 15–17, 21–3, 52, 54. *See also* meat, milk, standard of living
diphtheria, 34; and milk, 155
Diploma of Public Health (DPH), 76, 134
disease: animal, 3–4, 12, 20–51, 92, 138, 146, 154–9, 175–6, 186–91; endemic, 3, 12; epidemic, 72, 85, 87, 155–6, 176, 187; epizootic, 3, 4, 19, 21, 23, 25, 27–8, 32, 35, 38–9, 43, 53, 90, 92, 101–2, 106, 138, 162, 175–6, 186; food-borne, 1–2, 4, 9, 12, 20–9, 35–6, 38–9, 43, 52–63, 92–131, 133–6, 146, 153–61, 164, 175, 185–91; hereditary, 13–14, 31–4, 36, 38, 41–2, 49–50, 105, 184, 186; infectious, 1, 3, 23, 31–2, 34–6, 38–9, 45, 58–62, 129, 131, 155, 166, 188; and localised v. generalised infection, 58–62, 68, 81, 84–5, 89–92, 96–7, 141–2, 158, 185; milk-borne, 5, 9, 11, 12, 92, 106, 111–14, 116–17, 124–5, 128, 154–9, 164, 166, 185–7, 190; models, 25, 27, 31–3, 35–6, 39, 41, 58–62, 141; parasitic, 38, 42, 55; water-borne, 155; zoonotic, 1–3, 6–7, 25, 30, 33, 37, 39, 42, 52, 54–5, 58, 104, 121, 131, 148, 153, 156, 186–7; zymotic, 35, 42, 155. *See also individual diseases*
doctors, *see* medical profession
Dumbarton, 140
Durham, 180
Eastwood, Arthur, 121, 126, 129
Edinburgh, 22, 24, 25, 36, 37, 66, 85, 93, 139, 142
Edinburgh, University of, 22
Egypt, 118
Elliot, Thomas, 180
empire, health of, 12, 156
Environment, Food and Rural Affairs Committee, 5
epidemics, 72, 85, 87, 155–6, 176, 187
epidemiology, 12, 27, 101, 103, 114, 119, 128, 139, 155, 159
epizootics, 25, 27–8, 30, 36–9, 52, 92, 115, 138, 146, 178; and Board of Agriculture, 101–2, 176, 180–1; compensation for, 90, 101–2, 106, 176, 180–1; eradication of, 4, 40, 49–50, 90, 101–2, 106, 175–6, 179, 181; and farmers, 17, 19, 27–8, 38, 49–50, 90, 101–2, 106, 162, 175–6, 183–5; levels of, 3–4, 17, 19, 21, 23, 32, 35, 90, 92, 162, 175–6, 186; and LGB, 43; and medical profession, 23, 27–9, 35, 43; and Privy Council, 27, 29, 35, 49–50, 92, 176–7; and veterinarians, 25, 32, 36–40, 138. *See also individual diseases*, cattle plague, Contagious Diseases (Animals) Acts
Epsom, 161
Essex, 119
experiments: bacteriological, 3, 8, 47–8, 103–5, 109, 114–15, 119–22, 123–6, 128; and bovine tuberculosis, 9, 32–4, 37, 44, 46–9, 56, 97–8, 103–6, 114–15, 119–24, 127–9; DIY nature of, 8, 48, 121; inoculation, 22, 32–4, 37, 44, 47–8, 103–4, 126; for royal commissions on tuberculosis, 8, 103–6, 119–22, 126; and state, 8–9, 27, 35, 42, 103–6, 119–22, 128–9; transmission, 22, 32–4, 53–4, 56, 103, 120–1; and veterinary medicine, 36, 38–9, 53–4
expertise: butchers, 75, 79–80, 84, 97, 99–100, 107–8, 134–5, 189; and meat inspection, 8–9, 75, 78–81, 84–5, 94, 96, 133–42, 190; medical, 76, 79, 99, 139–40, 147; and milk, 161–2; in public health, 27, 39–40, 75–80, 84–5, 97, 133–40, 147, 161–2, 189–90; and sanitary inspectors, 77–9, 85, 133–5; and state, 102–3, 119–20; veterinarian, 24, 36, 39–40, 44, 71, 75, 94–5, 97, 99, 136–40, 147, 161–2, 168, 189–90

farcy, 32
farmers: and agricultural depression, 63, 166, 178–9, 184; and Board of Agriculture, 101–2, 166, 172–3, 181, 184–5; and bovine tuberculosis, 2, 49–50, 63–4, 68, 90, 100–2, 106–9, 115, 125, 158, 160–1, 165–6, 168, 170–1, 176, 178–9, 183–5, 189; and BSE, 1–2; and compensation, 90, 101–2, 106–9, 173, 176, 180–5; and Contagious Diseases (Animals) Acts, 50, 101–2, 106–8, 176, 184; dairy, 154, 158, 160–1, 164–5, 184; and epizootics, 17, 19, 27–8, 38, 49–50, 90, 101–2, 106, 162, 175–6, 183–5; influence of, 87, 101–2, 144, 154, 164–6, 176, 181, 183–5, 189; and LGB, 166, 180–1, 183–5; livestock, 14, 17, 63, 184; and local government, 87, 144, 165, 166, 180, 189; and milk, 154, 158, 160–1, 163–8, 170–3, 184; and public health, 2, 63–4, 87, 101–2, 144, 158, 160–1, 164–6, 170–1, 182, 184, 189; and tuberculin, 178–9, 184–5; and veterinarians, 63, 164, 184
Farrow, John, 84–5
Faulkner, Edwin, 56
First World War, 74, 132, 175, 185, 186, 191; and milk, 173, 183
Fleming, George, 38, 44; and bovine tuberculosis, 37, 45, 59, 65, 89; on contagion, 37; and public health, 37
food: adulteration, 3, 23, 26, 72–3, 77, 79, 84, 153–4, 156, 159–60, 171, 174; borne infection, 1–2, 4, 9, 12, 20–9, 35–6, 38–9, 43, 52–63, 92–131, 133–6, 146, 153–61, 164, 175, 185–91; consumption, 14–20, 23, 52, 154, 160–1, 176; hygiene, 101, 153–5, 159–60, 188; inspectors, 72, 74, 77–80, 131, 133; poisoning, 24, 29, 55; safety, 1–4, 13–14, 23, 26, 28, 35–9, 43, 51–2, 55–62, 64–9, 70–1, 73, 92–7, 99, 101–2, 112, 117, 125, 128, 131, 135–6, 141, 146, 161, 166, 171–3, 180–1, 183, 185–91; scares, 1–2, 4, 12, 27–9, 51, 92, 95, 112, 132, 145–6, 165, 185, 188, 190–1. *See also* BSE, CJD, meat, milk, sausages
foot-and-mouth disease, 17, 38, 39, 42, 52, 55, 176
Fox, Wilson, 34
France, 115, 123, 168, 178; veterinary medicine in, 32–4, 53, 178
fraud, 23, 52. *See also* adulteration
free trade, 125, 146, 173–4
Fyfe, Peter, 96–7

Gamgee, John: on diseased meat, 17, 25–7; and eradication of epizootics, 25, 27; opposition to, 25, 27; and rinderpest, 27
Gamgee, Joseph Sampson: on diseased meat, 17, 25; opposition to, 25
Garnett, Frank, 137
General Board of Health, 187
Gerlach, Andreas: and meat, 53; and tuberculosis, 34, 36, 44, 53; and veterinarians, 34, 36, 53
germ theory: and animal disease, 7, 28, 30–1, 35, 39, 42, 46–51, 56, 93, 99, 119, 187; and bovine tuberculosis, 7, 13, 30–1, 35, 39, 46–51, 56, 90, 99, 187; and contagion, 7, 28, 31, 35, 39, 42, 46–50, 56, 102, 119, 187; impact of on medicine, 7, 13, 48–51, 102, 124; and medical profession, 30, 35, 42, 47–51, 56, 120, 124, 187; and seed and soil, 49, 50, 65; and tuberculosis, 7, 13, 31, 35, 46–51, 65, 124, 187; and veterinarians, 35, 39, 48–9, 56, 90. *See also* bacteriology, Robert Koch
Germany, 123; and bovine tuberculosis, 31, 34–5, 46–7, 53, 114–15, 123–4, 128; Imperial Health Commission, 46, 128; and Koch, 46–9, 114–15, 128; meat inspection in, 115, 149; pathology in, 31, 34, 35, 53, 114; science and, 114–15, 123, 127; slaughterhouses in, 149; and tuberculosis, 31, 34–5, 46–7, 53, 114–15, 124, 128; and veterinary medicine, 34, 53, 114
Gibson, Thomas, 165
glanders, 30, 32, 34, 38, 39, 42, 47, 71
Glasgow: butchers in, 89, 94–7; diseased meat in, 8, 19, 68, 89, 93–7, 99–100, 144, 189; meat inspection in, 8, 89, 93–7, 99–100, 138–9, 144; police in, 92–6, 138; public health in, 93–7, 99–100, 138–9; slaughterhouses, 89, 96; trial, 8, 68, 91–102, 128, 132, 150, 189. *See also* Hugh Couper, James McCall, Charles Moore, James Russell
Glasgow United Fleshers' Society, 94
Glasgow Veterinary College, 24, 65, 94
Greenock, 88
Greenwich, 142

Griffith, Arthur Stanley, 129
grocers, 166–7, 170
Guildford, 169
Guy's Hospital, 136

Hamilton, David, 122, 126
Hardwicke, William, 56
Haywards Heath, 89
hereditary, 13–14, 31–4, 36, 38, 41–2, 49–50, 105, 184, 186. *See also* tuberculosis
Higgins, James, 19
Hill, Henry, 64
histology, 35, 103, 121, 126
Holborn, 132, 169
Holland, 178
Honourable Company of Butchers, 21, 71
horses, 24, 30, 32, 34, 38, 39, 40, 42, 47, 71, 138
hospitals, 25, 43, 44, 54; laboratories in, 42, 103, 119, 136, 169
housing: conditions, 66–7, 87, 163; overcrowded, 67, 72, 163
Huddersfield, 134
Hull, 86
Hunter, John, 24
hygiene, 23, 39; cowsheds and, 154–5; food, 101, 153–5, 159–60, 188

Imperial Department of Health, 46, 128
imported, cattle, 27, 28, 145–7
Independent Labour Party, 159
infants: and bovine tuberculosis, 11–12, 14, 66, 153, 156–7, 159–61, 169; health of, 5, 9, 12, 14, 153, 156–7, 159–61, 186; and milk, 9, 153, 156–7, 159–61, 169; mortality, 12, 48, 156, 159, 160–1, 169
inoculation: and animal disease, 22–3, 33, 103; experiments, 22, 32–4, 37, 44, 47–8, 103–4, 126; and royal commission on tuberculosis, 103–4, 126
inspection, *see* meat, milk, MOsH, sanitary inspectors, veterinarians
inspectors, *see* food inspectors, meat inspection, milk inspection, MOsH, sanitary inspectors, veterinarians
insurance, 90, 109–10
Interdepartmental Committee on Physical Deterioration, 156
International Veterinary Congress (1885), 57, 59
Islington, 28
Italy, 123

Japan, 123
Jenner Institute, 169
juries, butchers', 84

Kanthack, Alfredo, 101
Kay, James, 87
Kensington, 160 n43
King, James, 162, 168–9
Klebs, Edwin, 34, 53
Klein, Edward: on animal disease, 43, 44, 61; and bovine tuberculosis, 61; and contagion, 43, 61; and meat, 61
Koch, Robert: and 1901 congress, 8, 99, 112–19, 123, 126, 141; and anthrax, 46, 47, 60; and bacteriology, 7, 13, 31, 46–51, 56, 103, 114, 124–6; and bovine tuberculosis, 7, 13, 31, 46–51, 56, 58–9, 61, 99, 112–19, 122, 124; and cholera, 118–19; controversy surrounding, 8, 99, 112–19, 122, 124, 141; on diseased meat, 56, 59, 61, 114–15; and germ theory, 7, 13, 46–51, 56, 99, 124; and German medicine, 46–9, 114–15, 128; and laboratory innovations, 46, 48, 114; opposition to, 8, 99, 112–19, 122, 124, 141; public health, 7, 8, 48, 49, 56; public perception of, 115–16, 127–8; and royal commissions on tuberculosis, 8, 104, 112–13, 118–28, 130; status of, 7, 8, 47–9, 51, 99, 112–19, 124, 125–6; and tuberculin, 115; and tuberculosis, 7, 13, 46–9, 56, 99, 113, 124

laboratory, 8, 32, 103–6, 119, 135–6, 154, 169; accommodation, 42, 119–20, 121, 128–9, 136; and animal disease, 8, 36, 42, 60, 101, 103–6, 112, 118–24, 128–9, 135–6, 168, 169; bacteriological, 48, 101, 103, 119–22, 128–9, 135–6, 169; and bovine tuberculosis, 8, 36, 48, 101, 103–6, 112, 118–24, 128–9, 135–6, 168, 169; diagnostic work, 101, 119, 121, 128–9, 135–6, 168, 169; hospital, 42, 103, 119, 136, 169; and LGB, 42, 128–9; Manchester, 57, 168–9; for public health, 103, 119, 128–9, 135–6, 168–9; research, 7–8, 36, 42, 48, 60, 103–6, 112, 118–24, 128–30, 168; and royal commissions on tuberculosis, 8, 103–6,

112, 118–24, 128–9; Royalcot, 119, 120, 121; Stansted, 119, 128–9; university, 119, 168, 169. *See also* bacteriology
Laënnec, Rene, 31, 32, 46
Lancashire, 108
Lawson, John, 36
Leeds, 19, 87, 134, 164
Legge, T. Morrison, 84
Leicester, 56, 75
Letheby, Henry, 73, 87
Liebig, Julius, 22
Littlejohn, Arthur, 98
Liverpool, 13, 75, 79–80, 85, 100, 122, 139, 142, 161
livestock, *see* cattle, farmers, sheep, pigs
Local Government Board (LGB): and animal disease, 42–3, 48, 109, 119, 125, 141, 146, 168, 183, 188; and Board of Agriculture, 161–2, 166, 172–3, 180–1, 183, 185, 188; and bovine tuberculosis, 101–2, 117, 125, 128, 133, 140–3, 146–7, 168, 180–5, 188; and compensation, 109, 110, 125, 173, 180, 184–5; and dairies, 161, 166, 168, 171–3; and diseased meat, 101, 106, 117, 125, 127, 141–3, 146, 150, 157, 188; and farmers, 166, 180–1, 183–5; and food safety, 101–2, 117, 125, 141, 146, 161, 166, 168, 171–3, 180–1, 183, 188; Foods Section at, 125, 146, 183; and Glasgow trial, 101; and imported meat, 146–7; and laboratory, 42, 128–9; and meat inspection, 78, 117, 125, 133, 140–3, 146–7, 188; and milk, 125, 166, 168, 171–3, 180, 183, 188; and public health, 42, 78, 117, 125, 128–9, 132, 143, 146–7, 166, 168, 171–3, 183, 188; and royal commissions on tuberculosis, 101–3, 117–18, 125, 127–9, 140–1, 181; and sanitary inspection, 78, 133; and Tuberculosis Order, 173, 182–3; and veterinarians, 161
London: diseased meat in, 16, 18–19, 21, 25–6, 28, 71, 73, 80, 87, 132, 142, 146, 148; markets, 18–19, 21, 25–6, 28, 73, 132, 134; meat inspection in, 21, 25–6, 28, 71, 73, 78, 80, 85, 87, 132, 135, 142, 146, 148; Port of, 146; public health in, 25–6, 71, 73, 75, 78, 85, 87, 132, 135, 142, 146; university, 25. *See also* Holborn, Islington, City of London, Corporation of London, LCC
London, City of, 18, 80, 135
London, Corporation of, 73
London County Council (LCC), 19
Luton, 170
Lyon, 53, 123

McCall, James, 65, 94–5, 96
MacFadyean, John: and bacteriology, 61, 103–4; and bovine tuberculosis, 13, 61, 103–5, 115–16, 118, 127; on Koch, 115–16; and pathology, 61, 103–5; and royal commissions on tuberculosis, 103–5, 118, 127
magistrates: and meat, 83–5, 97–9, 104, 136, 144–5, 190; and milk, 170, 190; and MOsH, 83–5, 97–9, 144–5, 170, 190; and public health, 83–5, 97–9, 136, 144–5, 170, 190
Manchester, 13, 36, 77 n26, 83; butchers in, 57, 151; dairies in, 162, 168–9, 171; laboratory, 57, 168–9; milk regulation in, 168–9, 171; slaughterhouses in, 18, 151; Victoria University, 168, 169, 176
markets: and diseased meat, 16–21, 25, 52–3, 72, 188; Glasgow, 93–5; Holborn, 132; inspection of, 25, 72, 79, 85, 94; Islington, 28; in London, 18–19, 21, 25–6, 28, 73, 132, 134; Metropolitan Cattle, 28; Newgate, 18; poor and, 16, 18–20, 52; regulation of, 25, 72, 93, 94; West Smithfield, 134
Martin, Sidney: and bovine tuberculosis, 103–5, 118; and Koch, 118; and royal commissions on tuberculosis, 103–5, 118
meat: and bovine tuberculosis, 3–9, 11–12, 19, 29–30, 36–9, 43, 51–69, 70–1, 74–5, 81–5, 87–112, 117–18, 125–8, 131–52, 154, 157, 179, 185–91; and BSE, 1–2, 190–1; and CJD, 1–2; consumption of, 6, 15–20, 23, 52, 54; cooking of, 64–8, 154, 179; cultural value of, 15–16; in diets, 6, 15–17, 21–3, 52, 54; and food scares, 1–2, 12, 27–9, 51, 92, 112, 145–6, 187; frozen, 3, 147; imported, 3, 12, 28, 142, 145–7; inspection of, 7–9, 12, 25, 58, 68, 70–91, 93–7, 105–9, 110–11, 117, 128, 130–52, 154, 179–80, 187–90; market for, 15–16, 18, 188; prices, 15–17, 20, 23, 27, 54, 63, 151, 176; putrid, 71; supply, 15–20, 54, 63, 145–6; trade, 3, 16–20, 23, 25–6, 29, 52–4, 56, 63–4, 71, 74, 87–90, 92–3, 95–6, 99, 100–1,

105–8, 110, 117, 131, 141, 176, 186, 188; unwholesome, 6, 20, 26, 28, 43, 71, 84, 186. *See also* diseased meat, Glasgow, meat inspection, sausages, slaughterhouses
meat, diseased: in America, 3, 123–4, 145–6; appearance of, 43, 81–5, 105, 135, 142–4; and Board of Agriculture, 101, 188; bovine tuberculosis and, 3–9, 11–12, 19, 29–30, 36–9, 43, 51–69, 70–1, 74–5, 81–5, 87–91, 112, 117–18, 125–8, 131–52, 154, 157, 179, 185–91; and butchers, 19–20, 54, 56, 63–4, 71, 82–4, 87–90, 94–7, 99–101, 105–8, 110, 117, 134–5, 139, 141–4, 148–52, 180, 188–9; consumption of, 16–20, 23, 52, 54; and contagion, 1–3, 6–7, 14, 25, 28–9, 35–8, 43, 52–69, 96–106, 109–10, 131, 141–2, 157, 186–7; cooking and, 64–8, 154, 179; in diets, 6, 16–23, 52, 54; in Edinburgh, 25; and fraud, 23, 52; in Glasgow, 8, 19, 68, 89, 93–7, 99–100, 144, 189; and health, 2–4, 6–9, 14, 19–29, 35–8, 43, 52–79, 92–106, 109–10, 114–15, 117, 123–4, 128, 130–1, 141–2, 146, 186–91; identification of, 58–62, 81–5, 88–90, 96, 135–6, 141–2, 146–7, 186–7; imported, 3, 12, 28, 145–7; and inspection, 7–9, 12, 58, 68, 70–91, 93–8, 104–11, 117, 128, 130–52, 154, 179–80, 187–91; and Koch, 56, 59, 61, 114–15; and LGB, 101, 106, 117, 125, 127, 141–3, 146, 150, 157, 188; and localised v. generalised infection, 58–62, 68, 81, 84–5, 89–92, 97–101, 105–6, 117–18, 141–2, 185; in London, 16, 18–19, 21, 25–6, 28, 71, 73, 80, 87, 132, 142, 146, 148; magistrates and, 83–5, 97–9, 104, 136, 144–5, 190; market for, 16–20, 52, 54, 95, 188; and medical profession, 4–9, 12, 20–9, 35–6, 51–62, 64–70, 73–6, 81–4, 94–101, 105, 109–10, 117, 131, 139–40, 147, 185–90; and MOsH, 6–9, 19–29, 43, 50–2, 55–62, 64–9, 73–6, 81–5, 90–1, 94–7, 100–1, 105, 107, 117, 131–2, 135–7, 139–45, 147–8, 186–90; and police, 79, 82, 94–6, 143; and Port Sanitary Authorities, 146–7; and Privy Council, 19, 27, 29, 57, 64, 66; prosecution for, 26–7, 29, 43, 56, 75, 87–90, 92–3, 96–7, 99–101, 107–8, 125, 141–2, 144–5, 185; and public health, 1–4, 6–9, 13–14, 19–29, 35–7, 43, 51–97, 99–101, 104, 107, 109–10, 112–17, 119, 125–6, 128–9, 131–52, 154, 185–91; and royal commissions on tuberculosis, 93, 102–12, 117–18, 125–8, 132; sale of, 16–20, 23, 25–6, 29, 43, 52, 54, 56, 93, 96–7, 99–100, 110, 113, 117, 144–5; seizure of, 26, 29, 53, 56, 64, 71, 82–3, 84, 87–90, 96, 100, 102, 105–7, 109, 125, 132–3, 142–5, 149, 189; in slaughterhouses, 18–19, 26, 72, 82, 95–6, 111, 144, 148–52, 154, 188; and veterinarians, 7, 24, 37, 38–9, 53–7, 59–60, 71, 84–5, 93–5, 99–100, 106, 108, 127, 131, 136–40, 147, 185–90; voluntary surrender of, 87–8, 96, 143, 144. *See also* BSE, meat inspection, meat trade, slaughterhouses
meat inspection: appointment of, 25, 79–80, 85, 95, 109, 132–6, 139–40; and bacteriology, 101, 135–6, 139, 187; and bovine tuberculosis, 8–9, 70–1, 74–5, 81–5, 88–91, 93–8, 104–11, 117, 128, 130–52, 154, 175, 179–80, 185, 187–91; and butchers, 64, 71, 79–80, 82–4, 87–90, 94, 100–1, 106–9, 117, 134–5, 139, 141–4, 148–52, 188–9; and compensation, 90, 106–9, 112, 125, 132, 142; and Contagious Diseases (Animals) Acts, 106–8, 138–9, 140; co-operation in, 82–3, 87–90, 96–7, 142–4, 188–9; development of, 8–9, 12, 25, 70–91, 95–7, 99–100, 107–8, 110, 132–48, 154, 180, 187–91; evasion of, 82–3, 88–9, 143–4; expertise in, 8–9, 75, 78–81, 84–5, 94, 96, 133–42, 190; in Germany, 115, 149; in Glasgow, 8, 89, 93–7, 99–100, 138–9, 144; guidelines on, 81, 108–9, 135–6, 140–2; of imported meat, 145–7; insurance against, 90, 109–10; and LGB, 78, 117, 125, 133, 140–3, 146–7, 188; in London, 21, 25–6, 28, 71, 73, 78, 80, 85, 87, 132, 135, 142, 146, 148; magistrates and, 83–5, 97–9, 136, 144–5; in markets, 25, 72, 79, 85; by MOsH, 6, 8–9, 12, 20–9, 43, 50, 52, 55–62, 64–9, 73–6, 80–8, 90–1, 94–7, 99–101, 105, 107, 110, 117, 132, 135–6, 139–45, 147–8, 150–2, 186–90; and National Federation of Butchers and Meat Traders' Associations, 90, 106–8, 110, 142; nature of, 8–9, 68,

70–91, 94–7, 99–102, 105, 107–9, 132–52, 189–90; opposition to, 87–90, 100, 107–8, 142–3, 189; and police, 78–9, 82, 94–6, 99, 138, 143; and Port Sanitary Authorities, 146–7; practices, 8–9, 70–91, 94–7, 99–100, 102, 105, 107–8, 135–6, 140–8; problems with, 8–9, 58–62, 75, 81–91, 94–7, 99–100, 106–8, 132–6, 143–5, 147–8, 179, 189–90; and the public, 70–1, 188; and royal commissions on tuberculosis, 93, 102, 105–12, 117–18, 125–8, 132–3, 137, 140–2, 150; rural, 8–9, 143–4, 189; and sanitary inspectors, 8–9, 25, 59, 76–85, 96, 104, 132–6, 142–5; and slaughterhouses, 25, 82, 85–6, 145, 148–52, 154; strategies for, 81–5, 96–7, 102, 105, 109, 135–6, 140–2, 146–8; systematic, 91, 96, 135, 144, 147–8; training in, 75–9, 81, 107–8, 133–9, 189; and Tuberculosis Order, 140; uniformity in, 91, 108–9, 132, 140–5, 147–8; by veterinarians, 8–9, 12, 59–60, 71, 84–5, 127, 131–2, 134, 136–40, 147, 187–90

meat trade: bovine tuberculosis and, 16–20, 52, 54, 56, 63–4, 87–90, 92–3, 95–6, 99, 100–1, 105–8, 110, 117, 131, 142–5, 176, 186, 188–9; and compensation, 90, 107–9, 125, 132, 142, 176; and diseased meat, 3, 16–20, 23, 25–6, 29, 52–4, 56, 63–4, 71, 74, 87–90, 92–3, 95–6, 99–101, 105–8, 110, 117, 131, 142–5, 176, 186, 188–9; ‘evils’ of, 3, 22–3, 25–6, 52, 54, 74, 85, 186

media: on bovine tuberculosis, 2, 50, 92, 95; and BSE, 1–2, 190–1; food scares and, 1–2, 92, 95

medical officers of health (MOsH): and animal disease, 20–9, 43, 54, 56–8, 60, 73–4, 83–4, 139, 147, 155–6, 158–61, 186–90; appointment of, 23, 25, 70, 73, 75, 85; and bacteriology, 48, 50, 56, 62, 74–5, 135–6, 139, 169–70, 187; and bovine tuberculosis, 6–9, 12, 52, 55–62, 64–9, 73–4, 76, 81–5, 90–1, 94–7, 100–1, 105, 107, 117, 132, 135–7, 140–8, 158, 160–1, 164–70, 173, 186–90; and contagion, 26, 27, 43, 48–50, 62; and cooking, 64–8; and cowsheds, 162–5; and dairies, 163–8, 170–1, 173, 189; and diseased meat, 6–9, 19–29, 43, 50–2, 55–62, 64–9, 73–6, 81–5, 90–1, 94–7, 100–1, 105, 107, 117, 131–2, 135–7, 139–45, 147–8, 186–90; guidelines for, 81, 108–9, 135–6, 140–2; and housing conditions, 66–7, 87, 163; and Koch, 48, 51; and localised v. generalised infection, 58, 60–2, 68, 84–5, 92, 96–7, 100–1, 105, 141–2; in London, 25–6, 73, 75, 132, 142, 146, 148; magistrates and, 83–5, 97–9, 144–5, 170, 190; and meat inspection, 6, 8–9, 12, 20–9, 43, 50, 52, 55–62, 64–9, 73–6, 80–8, 90–1, 94–7, 99–101, 105, 107, 110, 117, 132, 135–6, 139–45, 147–8, 150–2, 186–90; and milk, 9, 12, 155–6, 160–1, 163–71, 173, 187–90; and National Federation of Butchers and Meat Traders’ Associations, 142; and pasteurisation, 160; professionalisation of, 24–5, 70, 75–6, 85–7, 133, 135, 139–40, 189–90; and rural areas, 144, 164, 167; and sanitary inspectors, 9, 77, 79, 96, 133–4; status of, 24, 75–6, 85–7, 94–5, 139–40, 189–90; training of, 75–6, 81, 134–5, 139, 189; and Tuberculosis Order, 181–3; and veterinarians, 7, 41, 44, 61, 84–5, 136, 139–40, 147, 160–1, 185, 190

medical profession: on animal disease, 6–9, 19–29, 33–6, 41–5, 48–62, 64–9, 83–4, 139–40, 147, 186–7; and bacteriology, 46–9, 56, 62, 103–4, 129, 135–6, 139; and bovine tuberculosis, 7, 12, 33–6, 41–62, 64–9, 73, 83–4, 99–101, 109–10, 116–17, 130, 147, 176, 185–6; and contagion, 27–9, 33–6, 41–9, 54, 60–2, 187; and cooking of meat, 64–8; and diseased meat, 4–9, 12, 20–9, 35–6, 51–62, 64–70, 73–6, 81–4, 94–101, 105, 109–10, 117, 131, 139–40, 147, 185–90; and food safety, 23, 26, 28, 35, 43, 51–2, 55–62, 64–9, 73, 131, 185–7, 189–90; and germ theory, 30, 35, 42, 47–51, 56, 120, 124, 187; and Koch, 47–51, 56, 116–17; and localised v. generalised infection, 58, 60–2, 68, 84–5, 97–9, 158, 185; and milk, 12, 155–61, 163–8, 170–1, 173, 187; and tuberculosis, 33–6, 41, 45, 47–51; and veterinarians, 8, 24, 39–41, 43–4, 48, 55, 59–61, 97, 100, 112, 137, 139–40, 147, 161, 176, 185, 190

Medical Research Committee (MRC), 129

medicine: and bacteriology, 31, 46–51, 56, 62, 101, 103–4, 114, 119–22, 128–9, 135–6, 139, 187; comparative, 24, 30–6, 42–3; laboratory, 8, 32, 42, 48, 103–6, 112, 118–24, 128–9, 135–6; science and, 27, 32, 35, 42–3; and tubercle bacillus, 31, 46–51, 62, 98, 101, 103, 113, 120, 126, 135–6, 187; veterinary, 24, 32, 39–40
Merthyr Tydfil, 73, 88
Metropolitan Association of Medical Officers of Health (MAMOH), *see* Society of Medical Officers of Health
Metropolitan Cattle Market, 28
microscopy, 44, 60, 103, 121, 181
Middlesbrough, 87, 151
Midlothian, 179
milk: adulteration, 153–4, 156, 159–60, 171, 174; bacteriological examination of, 154, 157, 167, 168–70, 187; and Board of Agriculture, 161–2, 166, 171–2, 181–3, 188; and bovine tuberculosis, 5, 9, 11–12, 92, 106–8, 111–14, 116–17, 124–5, 128, 130, 153–74, 176, 179, 186–90; and child health, 5, 9, 12, 153–4, 156–7, 159–60, 169, 172; clean, 159–61; and contagion, 5, 9, 11–12, 92, 106–8, 111–14, 116–17, 124–5, 128, 154–9; consumption of, 154, 160–1, 174, 176; and degeneration, 9, 108, 153, 156, 159, 175, 186; and diphtheria, 155; and disease, 5, 9, 11–12, 92, 106–8, 111–14, 116–17, 124–5, 128, 154–9, 164, 166, 185–7, 190; farmers and, 154, 158, 160–1, 163–8, 170–3, 184; and health, 5, 9, 11–12, 92, 106–8, 154–60, 165, 171–4, 187; inspection of, 92, 106–8, 111–14, 116–17, 124–5, 128, 130, 161–73, 175, 180, 185, 187–90; and Interdepartmental Committee on Physical Deterioration, 156; legislation, 166, 171–3, 180; and LGB, 125, 166, 168, 171–3, 180, 183, 188; in Manchester, 168–9, 171; and medical profession, 12, 155–61, 163–5, 166–8, 170–1, 173, 187; and MOsH, 9, 12, 155–6, 160–1, 163–71, 173, 187–90; municipal, 159–60; and pasteurisation, 5, 153, 160–1, 174, 186, 190; prices, 154, 160, 164, 166, 170, 172, 179; production, 154–6, 159–64, 166–7, 169, 170, 172, 179; and public, 153, 156–7, 159–61, 165, 188; and public health, 8–9, 12, 111, 119, 154–74, 187–90; pure, 113, 120, 159–61; regulation of, 9, 111, 113, 117, 125, 128, 153–4, 159–74, 187–90; and royal commissions on tuberculosis, 106, 108, 111–12, 125, 128, 130, 157, 160–1, 167, 172; from rural areas, 162, 164–7, 170, 172–3; and scarlet fever, 155–6; and sterilisation, 159–60; supply, 153–5, 160–1, 166–7, 170, 172; and typhoid, 155; and veterinarians, 8, 161–5, 167–9, 171–3, 187–90. *See also* dairy industry, National Clean Milk Society, National Health Society
milk inspection: and bacteriology, 154, 168–70; and bovine tuberculosis, 8–9, 111–13, 152–75, 180–1, 185, 187–90; co-operation in, 164–6, 168, 170–2; and cowsheds, 162–73, 188; and dairy industry, 113, 161, 166, 168, 170, 172; development of, 111–12, 154, 161–73, 180, 187–90; evasion of, 166, 168, 170–1; expertise in, 161–2; and farmers, 161, 164, 166, 168; limitations of, 153–4, 166–8, 170–1, 179, 187–90; in Manchester, 168–9, 171; by MOsH, 166–71, 187–90; nature of, 162–73, 187–90; opposition to, 168, 170–1, 189; and royal commissions on tuberculosis, 111–12, 161; in rural areas, 164, 167; in Sheffield, 169; strategies for, 168–70; and Tuberculosis Order, 161, 173, 181; by veterinarians, 8, 161–5, 167–9, 171–3, 187–90
Mill Road Infirmary, Liverpool, 122
Ministry of Agriculture, Fisheries and Food, 2
Moore, Charles, 96–7
mortality: infant, 12, 48, 156, 159–61, 169; tuberculosis, 5, 11–12, 14, 43, 116, 131, 156, 159, 168, 175, 189
municipal: milk supply, 159–60; slaughterhouses, 111, 149–52, 154, 185, 188, 190

National Association for the Prevention of Consumption: and bovine tuberculosis, 110, 113, 159; and milk, 110, 159
National Clean Milk Society, 160. *See also* pasteurisation, sterilisation
National Federation of Butchers and Meat Traders' Associations: and bovine tuberculosis, 90, 106–8, 110; and

compensation, 90, 106–8, 110, 142; and meat inspection, 90, 106–8, 110, 142; and royal commissions on tuberculosis, 106–8, 110; and slaughterhouses, 151
National Health Service, 190
National Health Society, 147, 160, 171. *See also* pasteurisation, sterilisation
National League for Physical Education and Improvement, 160, 171. *See also* pasteurisation, sterilisation
National Veterinary Association, 41, 116
New Edinburgh Veterinary College, 24
Newcastle, 63, 86, 142
Newgate market, 18
Newman, George, 15
Newsholme, Arthur, 57, 182; and meat, 62, 66, 88
Niven, James, 83, 168–9, 171
Northumberland and Durham Dairy and Tenant Farmers' Association, 180
Norway, 178
nuisance inspectors, *see* sanitary inspectors
nuisances: and cowsheds, 162–4; definition of 72, 77, 149; and public health, 8, 26, 70, 72, 77, 80, 146, 148–50, 162–4, 187, 189; and slaughterhouses, 72, 148–50

Oldham, 19, 168, 170
Ostertag, Robert, 124, 135
overcrowding, 67, 72, 163

Paddington, 26
Paisley, 89, 95
Paris, 47, 50, 157; congress, 33, 50, 57, 92, 97
Parkes, Edmund, 54, 64
Pasteur, Louis, 32
pasteurisation: and bovine tuberculosis, 5, 153, 160–1, 174, 186, 190; opposition to, 160
pathology: cellular, 31, 34–5; comparative, 6–7, 27, 29–39, 42–3, 53, 103, 119; of contagion, 27, 29–36, 42–3, 53, 114, 187; European, 6, 30–6, 39, 53, 59, 114, 157; experimental, 27, 33–6, 53, 103, 112, 119, 187; localised, 31, 35, 59–60; and royal commission on tuberculosis, 103–6, 109–10, 112, 119–22; of tuberculosis, 7, 30–6, 59–60, 83, 96–106, 109–10, 112, 114, 119–22, 157, 187; and veterinarians, 30, 36–9, 53–4
Pennsylvania, 123
phthisis, *see* tuberculosis
pigs, 25, 27, 29, 39, 90
pleuro-pneumonia, 17, 23, 26, 34, 43, 52, 54, 90, 100, 176, 181; and bovine tuberculosis, 38–9, 49–50, 96–7; contagion and, 25, 30, 33, 38–9, 43, 163; departmental committee on, 49–50, 56, 61, 65, 92, 96– 7, 102, 177, 183; and Privy Council, 49–50, 57, 92, 176; veterinarians and, 25, 33, 38–9
poisoning, food, 24, 29, 55
police: in Glasgow, 92–6, 138; and meat inspection, 78–9, 82, 94–6, 99, 138, 143; and public health, 78–9, 82, 92–6, 99, 138, 143; and sanitary inspectors, 79, 82, 96
polio, cattle, 10
poor: diets of, 15–17, 21–3, 54; and diseased meat, 6, 16–21, 52, 54; living conditions, 66–7
Port Sanitary Authorities, 146–7
ports: diseased meat at, 145–7; inspection of, 146–7
Portsmouth, 62, 89, 134
post-mortems, 12, 121, 159
Prague, 34, 53
Preston, 60
Privy Council: and animal disease, 35, 49–50, 64, 92, 104–5, 176–7, 183; and bovine tuberculosis, 49–50, 57, 92; and diseased meat, 19, 27, 29, 57, 64, 66; and epizootics, 27, 29, 35, 49–50, 92, 176–7; and pleuro-pneumonia, 49–50, 57, 92, 176; science and, 27, 29, 35, 104–5; veterinary department, 104, 176–7, 183. *See also* John Simon
professionalisation: of public health, 24–5, 70, 75–6, 85–7, 133–5, 139–40, 142, 189–90; of sanitary inspection, 77–80, 85, 133–5; of veterinary medicine, 24, 39–40, 136–40, 190
prosecution: avoidance of, 88–90, 107, 142, 170–1; and compensation, 90, 125, 142; for diseased meat, 26–7, 29, 43, 56, 75, 87–80, 90, 92–3, 96–7, 99–101, 107–8, 125, 141–2, 144–5, 185; for diseased milk, 170–1, 185; in Glasgow, 93, 96–7
ptomaines, 24
public: and bovine tuberculosis, 2, 52, 63, 92, 99–100, 108, 110, 112, 116, 127, 132, 160–1, 165, 181–2, 188; and BSE, 1–2, 190–1; and food scares, 1–2, 28,

112, 132, 165, 188; and Koch, 115–16, 127–8; and meat, 28, 52, 70–1, 92, 99, 112, 132, 188; and milk, 153, 156–7, 159–61, 165, 188; and tuberculosis, 63, 110, 116
public health: bacteriology in, 7, 47–51, 56, 62, 74–5, 103, 119, 128–9, 135–6, 139, 167, 169–70, 187; and bovine tuberculosis, 3–4, 6–9, 14, 29, 43–5, 47–52, 55–62, 64–9, 73–4, 76, 81–5, 90–101, 104, 107, 109–10, 112–17, 119, 126–9, 131–52, 163–73, 185–91; co-operation in, 87–8, 143–5, 168, 170, 189; and cowsheds, 3, 13, 55, 162–73, 188; and dairies, 13, 113, 161, 163–73, 188; development of, 8, 41, 70–2, 74, 85, 103, 109, 127, 132, 142, 146–7, 164–73, 176, 187–90; and diseased meat, 1–4, 6–9, 13–14, 19–29, 35–7, 43, 51–97, 99–101, 104, 107, 109–10, 112–17, 119, 125–6, 128–9, 131–52, 154, 185–91; expertise in, 27, 39–40, 75–80, 84–5, 97, 133–40, 147, 161–2, 189–90; and farmers, 2, 63–4, 87, 101–2, 144, 158, 160–1, 164–6, 170–1, 182, 184, 189; and food safety, 1–4, 13–14, 23, 26, 28, 35, 37, 39, 43, 51, 70–1, 92–7, 99, 112, 117, 125, 128, 131, 141, 146, 187–91; in Glasgow, 93–7, 99–100, 138–9; inclusive and exclusive measures, 74, 163, 176, 189; laboratories, 103, 119, 128–9, 135–6, 168–9; language of, 51, 163; legislation, 25, 29, 43, 72–4, 76–7, 81, 94–6, 132, 146–7, 150, 161, 163–71; and LGB, 42, 78, 117, 125, 128–9, 132, 143, 146–7, 166, 168, 171–3, 183, 188; limitations of, 41, 85–7, 93–7, 143–4, 164–5, 166–8, 170–2, 189–90; localism in, 8, 70, 73–4, 76–7, 85, 109, 131–2, 141–2, 166–7, 187–90; in London, 25–6, 71, 73, 75, 78, 85, 87, 132, 135, 142, 146; magistrates and, 83–5, 97–9, 136, 144–5, 170, 190; in Manchester, 168–9, 171; manuals, 51, 81, 135–6, 141; and milk, 8–9, 12, 111, 119, 154–74, 187–90; and nuisances, 8, 26, 70, 72, 77, 80, 146, 148–50, 162–4, 187, 189; pamphlets, 160, 170; police and, 78–9, 82, 92–6, 99, 138, 143; and Port Sanitary Authorities, 146–7; profession, 24, 25, 70, 75–6, 86–7, 103, 133–5, 139–40, 142, 189–90; and research, 27, 42, 119, 128–9; rural, 8–9, 87, 143–4, 164–7, 170, 172–3, 189; and science, 27, 29, 101, 119, 128–9; and slaughterhouses, 72, 82, 148–52, 154, 188; and veterinarians, 8–9, 37, 39–40, 55, 59–60, 71, 84–5, 131, 136–40, 147, 161–2, 168–9, 171–2, 186–90. *See also* DPH, LGB, meat inspection, milk inspection, MOsH, sanitary inspectors, John Simon
Pure Food and Health Society, 171

quarantine, 38, 118–19, 178

rabies, 3, 22, 23, 30, 32, 39, 42, 47, 71, 186
railway, 167, 168
Ravenel, Mazyck, 123–4
Raw, Nathan, 122
Rayer, Pierre-François-Olive, 32
Registrar General, 159
Reid, George, 134
Renfrew, 94
research: laboratory, 7–8, 36, 42, 48, 60, 103–6, 112, 118–24, 128–30, 168; and public health, 27, 42, 119, 128–9; royal commissions and, 8, 103–6, 112, 118–24, 128–9; state and, 8, 27–9, 35, 42, 101–6, 112, 118–24, 128–9; and veterinary medicine, 32, 37–9, 43, 103
rinderpest, 52, 53–4, 90, 100, 115, 163, 167; epizootic of, 17, 27–30, 35, 37–9, 146, 162, 176; and germ theories of disease, 28, 30, 39; medical profession's view of, 28, 35; royal commission on, 28–9, 35; veterinarians and, 37–9
Rotherham, 56
Royal Agriculture Society, 166
Royal College of Veterinary Surgeons (RCVS), 36, 40, 55
Royal Commission, cattle plague, 28–9, 35
Royal Commission, Scottish, on Physical Training, 67
Royal Commission, tuberculosis: appointment of, 93, 102, 107, 117–18; bacteriology and, 8, 103–5, 109, 119–26, 128; and Board of Agriculture, 101–2, 104, 106, 110, 181; British model and, 8, 93, 109, 112, 118–22, 125–8, 130; butchers and, 105–8, 110; compensation and, 106–10, 117, 125, 181; and diseased meat, 93, 102–12, 117–18, 125–8, 132; and eradication of bovine tuberculosis, 108–9, 175, 179, 181; experiments for, 8, 103–5, 119–22, 126; first (1890–5), 102–7,

109–10, 132, 137–8, 140, 157; international, 123–8; investigators for, 103, 118–22, 126, 128; and Koch, 8, 104, 112–13, 118–28, 130; laboratory and, 8, 103–6, 112, 118–24, 128–9; and LGB, 101–3, 117–18, 125, 127–9, 140–1, 181; and localised v. generalised infection, 105–6, 117–18, 140–1; and meat inspection, 3, 102, 105, 106–12, 117–18, 125–8, 132–3, 137, 140–2, 150; and milk, 106, 108, 111–12, 125, 128, 130, 157, 160–1, 167, 172; and National Federation of Butchers and Meat Traders' Associations, 106–8, 110; pathology and, 103–6, 109–10, 112, 119–22; problems in, 105, 121–2; reactions to, 106–7, 109–10, 126–8, 130; and research, 8, 103–6, 112, 118–24, 128–9; and science, 8, 103–7, 112, 118–24, 127–9; second (1895–8), 8, 80, 84, 93, 107–10, 113, 115, 117, 132–3, 137, 140–1, 150, 157, 160–1, 167, 175, 179; and slaughterhouses, 111, 150; support for, 106–10, 117, 126–8; third (1901–11), 8, 112, 117–30, 172, 181; veterinarians and, 106, 108, 127, 137; witnesses to, 102, 107–8, 167. *See also* John MacFadyean, Sidney Martin, Royalcot, Stansted, German Sims Woodhead
Royal Free Hospital, 25
Royal Statistical Society, 15
Royal Veterinary College (RVC), 13, 24, 40, 41, 98, 115
Royal Victoria Hospital, Netley, 54
Royalcot, 119, 120, 121
Runciman, Walter, 182, 183
rural: cowsheds, 164, 167; dairies, 162, 164–5, 167; meat inspection, 8–9, 143–4, 189; meat supply, 19, 143–4; milk inspection, 164, 167; milk supply, 162, 164–7, 170, 172–3; MOsH, 144, 164, 167; public health, 8–9, 87, 143–4, 164–7, 170, 172–3, 189; slaughterhouses, 144
Russell, James: and bovine tuberculosis, 94–7; and meat inspection, 94–7; and public health, 94–7, 139; and tuberculosis, 139

St Bartholomew's Hospital, 43, 169
Salford, 142
salmonella, 1
Samuel, Herbert, 173
sanatoriums, 5, 131, 186
Sanderson, John Burdon, 26; and tuberculosis, 33–4; on Villemin, 33–4
sanitary inspectors: appointment of, 25, 77–80, 85, 133–5; competence of, 59, 78–9, 133–5, 143; expertise and, 77–9, 85, 133–5; and LGB, 78, 133; limitations of, 77–80, 85, 133–5, 143; and meat, 8–9, 25, 59, 76–85, 96, 104, 132–6, 142–5; and MOsH, 9, 77, 79, 96, 133–4; and police, 79, 82, 96; professionalisation of, 77–80, 85, 133–5; status of, 77–80, 133–4; training, 78–9, 133–5, 189; and veterinarians, 59. *See also* meat inspection, milk inspection, public health, Sanitary Inspectors Association, Sanitary Inspectors' Examination Board, Sanitary Institute
Sanitary Inspectors Association, 133
Sanitary Inspectors' Examination Board, 133–4
Sanitary Institute, 78, 133–5
Saunders, Sedgwick, 18
sausages, 17; diseased, 25
Scarborough, 142
scarlet fever, 53; and milk, 155–6
science: and bovine tuberculosis, 3, 7–9, 30–3, 36, 47–51, 53–4, 101–30; in Britain, 8, 27, 42, 103, 112, 118–19, 127; endowment of, 103, 118–19, 128–9; in Germany, 114–15, 123, 127; and the laboratory, 8, 42, 101, 103–4, 107, 112, 118–29; and Privy Council, 27, 29, 35, 104–5; and public health, 27, 29, 101, 119, 128–9; routinised, 120; and royal commissions on tuberculosis, 8, 103–7, 112, 118–24, 127–9; and state, 8, 27–9, 42, 103–7, 112–13, 117–21, 128–9; value of, 8, 27, 42, 97–9, 101, 103, 107, 112, 118–19, 128–9; and veterinary medicine, 38–40, 103, 138. *See also* bacteriology, laboratory, pathology
Second World War, 185, 190
sheep, 27, 39
sheep-pox, 34, 44, 54
Sheffield, 64, 84, 139, 169; and milk inspection, 169
Sheffield Butchers' Association, 64
Shipley, W., 12
Short, A., 159

Simon, John: and meat, 27; and public health, 27, 35; and research, 27, 35, 42, 129; and zoonoses, 27, 35
Sinclair, Upton, 145–6
slaughterhouses: America, 145–6; and butchers, 82, 148–52; conditions in, 72, 145, 148–9; diseased meat and, 18–19, 26, 72, 82, 95–6, 111, 144, 148–52, 154, 188; in Germany, 149; in Glasgow, 89, 96; inspection of, 25, 82, 85–6, 91, 95–6, 135, 144–5, 147–52, 154; municipal, 111, 149–52, 154, 185, 188, 190; and National Federation of Butchers and Meat Traders' Associations, 151; and nuisances, 72, 148–50; and public health, 72, 82, 148–52, 154, 188; regulation of, 72, 95–6, 144, 148–52; and royal commissions on tuberculosis, 111, 150; rural, 144
slaughtering, practices, 72, 148
smallpox, 19, 28, 32, 34, 42, 60
Smith, Theobald, 114, 127
Society of Medical Officers of Health (SMOH): and diseased meat, 18, 26, 29, 42, 62, 73, 79, 110; and milk, 163; and sanitary inspectors, 79
Spencer, Earl of, 13
spongiform encephalopathies, 1–2. *See also* BSE
Staffordshire, 165
standard of living, 15–16, 154; and meat, 15
Stansted, 119, 128–9
state medicine, *see* public health
Stephenson, Clement, 63
Stepney, 148
sterilisation, milk: and bovine tuberculosis, 5, 153, 160–1, 173; opposition to, 160. *See also* milk, pasteurisation
Stockport, 188
Stockton, 88
Sunderland, 87
Swansea, 139, 150
Swansea Butchers' Association, 139
Sweden, 178
syphilis, 32

tariff reform, 146, 173
tea, 154
Thomson, Comin, 98
Treasury, 183
trichinous, 25, 29
Trotter, Alexander, 138–9
tubercle bacillus: and bacteriology, 7, 31, 46–51, 56, 62, 101, 103, 113–14, 120–2, 126, 128, 135–6, 187; identification of, 7, 31, 46–51, 56, 58, 90, 96, 98, 101, 103, 112–14, 116–17, 120–2, 126, 128, 135–6, 158, 187; Koch and, 7, 31, 46–51, 56, 58, 112–14, 116–17, 127; and medicine, 31, 46–51, 62, 98, 101, 103, 113, 120, 126, 135–6, 187; royal commissions and, 103, 112, 120–2, 125–8; and veterinarians, 48–9, 51
tuberculin, 104, 115; in Denmark, 177–8; in eradicating bovine tuberculosis, 176–81, 184–5; and farmers, 178–9, 184–5; and state, 180–1; support for, 179–81, 184–5; testing, 13, 176, 178–81, 184–5; and Tuberculosis Order, 184–5
tuberculosis: abdominal, 11; aetiology, 31, 32–4, 41–2, 47–51, 81, 114; artificial, 33–4; bacteriology and, 31, 46–51, 56, 62, 103–5, 109, 114–15, 119–26, 128, 187; in children, 3, 11–12, 14, 66, 124, 126, 153–4, 156–7, 159–60, 168, 172; and contagion, 3, 21–2, 31–4, 36, 41–51, 59–62, 96–106, 109–10, 112–30, 186–7; crusade against, 3–5, 14, 110–11, 113, 131, 159–60, 175, 179, 182, 186, 188; fear of, 49, 110, 116; germ theory and, 7, 13, 31, 35, 46–51, 65, 124, 187; hereditary, 13–14, 31–4, 36, 41–2, 49–50, 105, 115, 184, 186; historiography, 5, 11, 153; and Koch, 7, 13, 46–9, 56, 99, 113, 124; localised v. generalised, 58–62; and medical profession, 33–6, 41, 45, 47–51; mortality, 5, 11–12, 14, 43, 116, 159, 168; and MRC, 129; and pathology, 7, 30–6, 59–60, 83, 96–106, 109–10, 112, 114, 119–22, 157, 187; prevention, 5, 102, 104, 106–7, 109–11, 113, 115, 123, 127, 129, 131, 146–7, 153–4, 162–76, 182, 185–91; public and, 63, 110, 116, 188; pulmonary, 5, 11–14, 31–4, 41–2, 46, 98, 108, 113, 115–16, 124, 127, 129, 131, 186; and sanatoriums, 5, 131, 186; and seed and soil, 49–50, 65; social meanings of, 5, 116. *See also* bovine tuberculosis, National Association for the Prevention of Consumption

tuberculosis, bovine: bacteriology and, 3, 7, 13, 22, 31, 46–51, 56, 62, 103–5, 109, 114–15, 119–26, 128, 135–6, 154, 157, 169–70, 187–8; and Board of Agriculture, 101–2, 104, 106, 110, 161, 166, 171–3, 180–3, 185, 188; British model of, 8, 93, 109, 112, 118–22, 125–8; and butchers, 19, 50, 54, 56, 63–4, 82–3, 87–90, 94–7, 99–101, 105–8, 110, 117, 125, 131, 141–4, 148–52, 189; in cattle, 2–3, 12–14, 17, 19, 30–1, 33, 36–40, 52, 54, 56, 59–62, 82, 90, 96, 99, 102, 104–6, 110, 114–15, 125, 128, 145–7, 157–9, 162, 168, 170, 175, 177–8, 186–7; in children, 12, 14, 66, 124, 126; compensation for, 90, 100, 106–10, 125, 132, 142, 173, 180–5; and contagion, 3, 6–7, 13–14, 29, 30–49, 53–8, 60–2, 96–106, 109–10, 112–30, 154–9, 186–7; and Contagious Diseases (Animals) Act, 50, 90, 107–9, 138, 176; and cooking of meat, 64–8, 154, 179; cost of, 14, 92, 115, 179, 181, 184–5; and cowsheds, 3, 13, 55, 162–73, 188; and crusade against consumption, 3–4, 14, 110–11, 113, 159–60, 175, 179, 182, 186, 188; in dairies, 13, 157–9, 162–73, 188; and dairy industry, 113, 153–4, 158, 161–74, 179–80, 183, 189; and Denmark, 115, 177–8; and Departmental Committee on Pleuro-pneumonia, 50, 56, 61, 65, 92, 96–7, 102, 177, 183; and eradication of, 2, 4, 14, 92, 102, 104, 106–7, 109–10, 113, 123, 129, 131, 160, 175–86, 190; and farmers, 2, 49–50, 63–4, 68, 90, 100–2, 106–9, 115, 125, 158, 160–1, 165–6, 168, 170–1, 176, 178–9, 183–5, 189; fear of, 2–3, 12, 14, 92, 96, 110, 153, 186; and France, 32–4, 53, 115, 123, 178; and germ theory, 7, 13, 30–1, 35, 39, 46–51, 56, 90, 99, 187; and Germany, 31, 34–5, 46–7, 53, 114–15, 123–4, 128; and Glasgow trial, 8, 19, 68, 91–102, 128, 132, 150, 189; hereditary, 13–14, 31–4, 38, 49, 65, 105, 115, 184, 186; identification of, 12–13, 60, 68, 82–5, 88–90, 97–8, 102–5, 119–28, 176, 184, 186–7; and Koch, 7, 13, 31, 46–51, 56, 58–9, 61, 99, 112–19, 122, 124; levels of, 2, 11–12, 19, 110, 131, 188; and LGB, 101–2, 117, 125, 128, 133, 140–3, 146–7, 168, 180–5, 188; localised v. generalised, 58–62, 68, 81, 89–92, 96–101, 105–6, 117–18, 126, 140–2, 158, 185; in man, 2, 30–50, 59–62, 92–128, 131, 157–9, 175, 185–91; and meat, 3–9, 11–12, 19, 29–30, 36–9, 43, 51–71, 74–5, 81–5, 87–106, 111–12, 117–18, 125–8, 131–52, 154, 157, 179, 185–91; and meat inspection, 8–9, 70–1, 74–5, 81–5, 88–91, 93–8, 104–11, 117, 128, 130–52, 154, 175, 179–80, 185, 187–91; and meat trade, 16–20, 52, 54, 56, 63–4, 87–90, 92–3, 95–6, 99–101, 105–8, 110, 117, 131, 142–5, 176, 186, 188–9; and medical profession, 7, 12, 33–6, 41–62, 64–9, 73, 83–4, 99–101, 109–10, 116–17, 130, 147, 176, 185–6; and milk, 5, 8–9, 11–12, 92, 106–8, 111–14, 116–17, 124–5, 128, 130, 152–74, 176, 179–81, 185–90; morphology of, 44, 49, 82, 114, 116, 122, 126–7; mortality from, 11–12, 14, 131, 156, 159, 168, 175; and MOsH; 6–9, 12, 52, 55–62, 64–9, 73–4, 76, 81–5, 90–1, 94–7, 100–1, 105, 107, 117, 132, 135–7, 140–8, 158, 160–1, 164–70, 173, 186–90; and National Federation of Butchers and Meat Traders' Associations, 90, 106–8, 110; and pasteurisation of milk, 5, 153, 160–1, 174, 186, 190; and pleuro-pneumonia, 38–9, 49–50, 96–7; prevention of infection from, 102, 104, 106–7, 109–11, 113, 115, 123, 127, 129, 131, 146–7, 153–4, 162–76, 182, 185–91; and Privy Council, 49–50, 57, 92; public and, 2, 52, 63, 92, 99–100, 108, 110, 112, 116, 127, 132, 160–1, 165, 181–2, 188; and public health, 3–4, 6–9, 14, 29, 43–5, 47–52, 55–62, 64–9, 73–4, 76, 81–5, 90–101, 104, 107, 109–10, 112–17, 119, 126–9, 131–52, 163–73, 185–91; royal commissions on, 8, 80, 84, 93, 102–7, 109–10, 112, 117–30, 132–3, 137, 140–1, 142, 150, 157, 160–1, 167, 172, 175, 179, 181; in rural areas, 19, 144; science of, 3, 7–9, 30–3, 36, 47–51, 53–4, 101–30; and sterilisation of milk, 5, 153, 160–1, 173; symptoms, 12, 82–3; transmission of, 4, 11–12, 30–50, 52–64, 69, 102–30, 186–7; and

tuberculin, 176–81, 184–5; and Tuberculosis Order, 161, 173, 181–5; of the udder, 158, 168, 181; and veterinarians, 7, 12, 32–3, 36–41, 43–4, 47, 49–50, 53–64, 82, 84–5, 93–5, 97, 99–100, 105–6, 108, 127, 131, 136–40, 147, 161–2, 168, 176, 185–90; virulence of, 114–15, 121, 123, 126; as a zoonosis, 3, 6–7, 30–51, 54–5, 58, 63, 103–5, 112–31, 148, 153, 175, 186–7. *See also* meat, milk, National Association for the Prevention of Consumption, Royal Commission on tuberculosis, tuberculosis
typhoid, 32, 34, 45; and meat, 1, 29; and milk, 155

udder, tuberculosis of, 158, 168, 181
university: Aberdeen, 122; Cambridge, 44; Edinburgh, 22; laboratories, 119, 168, 169; Liverpool, 101, 161; London, 25; Victoria, 168, 169, 176
University College, Liverpool, 101
University College, London, 25
urbanisation, consequences of, 11, 70, 72, 74, 87

vaccine, 4
Vacher, Francis: and bovine tuberculosis, 19, 44, 45, 62; on meat inspection, 62, 88, 143
veterinarians: and animal disease, 1, 7–8, 24–5, 32–3, 36–41, 43–4, 48, 53, 55, 59–60, 71, 105, 131, 137–9, 147, 161–2, 168, 185–90; and bacteriology, 47–9, 103, 139; and bovine tuberculosis, 7, 12, 32–3, 36–41, 43–4, 47, 49–50, 53–64, 82, 84–5, 93–5, 97, 99–100, 105–6, 108, 127, 131, 136–40, 147, 161–2, 168, 176, 185–90; BSE and, 1; and comparative pathology, 30, 36–9, 53–4; on contagion, 24–5, 30, 32–3, 36–9, 43–4, 59–60, 138, 186; and Contagious Diseases (Animals) Acts, 40, 138–40, 161; and cooking of meat, 65–6; and cowsheds, 162–3, 164–5; and dairies, 160–1, 168, 172; and diseased meat, 7, 24, 37–9, 53–7, 59–60, 71, 84–5, 93–5, 99–100, 106, 108, 127, 131, 136–40, 147, 185–90; elite, 24, 36, 40, 55, 60, 176, 187–8; and epizootics, 25, 32, 36–40, 138; European, 24, 34, 36, 37, 39, 53–4, 55, 71, 136; and expertise, 24, 36, 39–40, 44, 71, 75, 94–5, 97, 99, 136–40, 147, 161–2, 168, 189–90; and farmers, 63, 164, 184; and germ theory, 35, 39, 48–9, 56, 90; and LGB, 161; and localised v. generalised infection, 58–62, 68, 84–5, 185; and meat inspection, 8–9, 12, 59–60, 71, 84–5, 127, 131–2, 134, 136–40, 147, 187–90; and medical profession, 8, 24, 39–41, 43–4, 48, 55, 59–61, 97, 100, 112, 137, 139–40, 147, 161, 176, 185, 190; and milk, 8, 161–5, 167–9, 171–3, 187–90; and MOsH, 7, 41, 44, 61, 84–5, 136, 139–40, 147, 160–1, 185, 190; and pleuro-pneumonia, 25, 33, 38–9; professionalisation of, 24, 39–40, 136–40, 190; and public health, 8, 9, 37, 39–40, 55, 59–60, 71, 84–5, 131, 136–40, 147, 161–2, 168–9, 171–2, 186–90; and research, 32, 37–9, 43, 103; and rinderpest, 37–9; and royal commissions on tuberculosis, 106, 108, 127, 137; and sanitary inspectors, 59; and science, 38–40, 103, 138; status, 24, 28, 39–41, 61, 132, 137–40, 172; training, 24, 36, 39–41, 134, 137–8, 140; and tubercle bacillus, 48–9, 51; and Tuberculosis Order, 140, 161. *See also* Association of Veterinary Officers of Health, George Fleming, John Gamgee, John MacFadyean, Royal Veterinary College (RVC)
veterinary medicine: in France, 32–4, 53, 178; in Germany, 34, 53, 114; profession, 24, 36, 39–41, 61, 103, 136–40, 190; and public health, 8, 9, 37, 39–40, 55, 59–60, 71, 84–5, 131, 136–40, 147, 161–2, 168–9, 171–2, 186–90; and research, 32, 37–9, 43, 103; and science, 38–9, 103, 138; status of, 24, 28, 39–41, 61, 132, 137–40, 172; training in, 24, 39–41, 134, 137–8, 140. *See also* George Fleming, John Gamgee, John MacFadyean, Royal Veterinary College (RVC), veterinarians
Victoria University, Manchester, 168, 169, 176
Villemin, Jean-Antoine: and bovine tuberculosis, 32–3, 36, 43, 46, 53; and tuberculosis, 32, 46, 53
Virchow, Rudolf, 34, 35, 53
vivisection, 43, 150; anti-, 150

Wakefield, 165
Wallace, Edgar, 16
Walley, Thomas, 41; and bovine tuberculosis, 37–8, 66; and cooking of meat, 66
Walpole farm, 119
Ward Richardson, Benjamin, 25
Warrington, 43
Washbourn, John, 136
Washington, 127, 129
West Hartlepool, 89
West Smithfield, 134
Whitelegge, Arthur, 65
Wigan, 100–1, 140
Williams, William, 59
Wilson, George, 81
Windermere, 136
Woodhead, German Sims: and bacteriology, 103–5; and Koch, 118; and royal commissions on tuberculosis, 103–5, 118, 127
World Health Organisation, 185

Yarmouth, 13
Yorkshire, 19
Young, McLauchlan, 122, 126
Young, William, 96

zoology, 42
zoonosis, 1–3, 6–7, 25, 30, 33, 37, 39, 42, 52, 54–5, 58, 63, 104, 121, 131, 148, 153, 156, 186–7. *See also* anthrax, bovine tuberculosis, BSE, rabies
zymotic disease, 35, 42, 155